EXOTIC VEGETABLES A~Z

Josephine Bacon

EXOTIC VEGETABLES A~Z

Line drawings by Soun Vannithone
Photographs by Julian Seaton

Salem House Publishers
TOPSFIELD, MASSACHUSETTS

DEDICATION: TO HANNA

ACKNOWLEDGEMENTS

Thanks are due to local vendors of exotic fruits and vegetables, especially Mr N. Patel of West Green Road, London, N.15, the Charalambides family, fruit importers of Caledonian Road, N.7, and Bob Milne of Seven Sisters Market. Many thanks also to my friend, Audrey Ellison, who helped me correctly identify several fruits and vegetables which are easily confused with each other, and to Jack Kessler, of Angmering, Sussex, who lent me some valuable old books on the subject. The author would also like to thank Dr and Mrs Dajani of Bet Hanina and Jericho, for the visit to their orchard, the New Zealand Apple and Pear Council, Dr Randy Keim of the South Coast Field Station, Santa Ana, California, U.S.A., and Avraham Shinar, Director of the Technical Assistance and Foreign Relations Department of the Ministry of Agriculture, Tel Aviv, Israel.

This edition first published in Great Britain by Xanadu Publications Ltd 1988

First published in the United States by Salem House Publishers, 1988, 462 Boston Street, Topsfield, MA 01983.

ISBN 0–88162–356–3

Manufactured in Great Britain

CONTENTS

INTRODUCTION

The astonishing number and variety of vegetables that are increasingly available in our shops and markets are perhaps due to three main trends. The first is that people now travel more widely than ever before and, while abroad, discover new and exciting foods that they then wish to use back home. Second, large immigrant and ethnic populations in the big cities create a demand for vegetables from their countries of origin, and their cuisines are in turn propagated and adopted by others. Third, books, magazines and TV have made us all much more sophisticated in our tastes.

This book, like its companion volume *Exotic Fruits A-Z*, offers guidance through this somewhat bewildering profusion by listing most of the exotics that you are likely to encounter and under each heading giving notes on what it is, where it comes from, when it's ripe/available/in good condition, and finally – and all importantly – what to do with it. The mere fact that a vegetable is classfied as 'exotic' (it's a vague term anyway) is not necessarily a sure recommendation of its tastiness. Some of the Indian vegetables, for instance, are decidedly bitter and while this may be concealed in hot curries it can emerge unpleasantly with milder spicing.

The recipes that are included point to some of the possibilities with these vegetables, but these are included only when they are unfamiliar enough to warrant it. For many items, the methods of preparation are the same as for our more familiar vegetables and need no special discussion; the huge variety of starchy vegetables that take the place of the potato in many tropical cuisines, for instance, offer fascinating variations on the familiar theme, but can usually be cooked much as potatoes are. Some cooking methods in the book are cross-referenced from one entry to another.

Overall, the word 'vegatable' is used in its broadest sense to describe plants that are principally *used* as vegetables even if some of them are technically fruits (e.g. breadfruit), and there are a few cases where something is employed as a vegetable in its unripe state and eaten as a fruit when it is ripe (e.g. mango and papaya): these will be found in *Exotic Fruits A-Z*. Nuts, mushrooms and those vegetables that are used primarily as spices are omitted, the first two because they are not quite vegetables in the present sense, and the last because herbs and spices need separate treatment and will, indeed, be the subjects of two further books in this series.

As with fruits, many of these vegetables are known by a bewildering number of names, and I apologize in advance if any local names have been omitted. But the plan of each entry – most common name, followed by the Latin name and then a list of alternative names – should cover most possibilities, and all the names are of course in the index: look there if you can't immediately find what you are looking for. The line drawings provide an instant visual check, and there are also photographs of some of the interesting vegetables included.

The decision as to what exactly is 'exotic' is in the end a personal one, as readers in different parts of the world will have their own ideas about the matter, but I hope that in this book I have covered most of the contenders. In the recipes, I have given Imperial measures first, then metric and finally American, to cover all variations in *that* department. An Imperial pint contains 20 fluid ounces, an American one just 16; otherwise, just choose the units you know best and stick with them. All spoon measures are level.

Bon appetit!

Ackee *Blighia sapida*

Akee

Although this plant originates from West Africa, it was introduced to the West Indies in the eighteenth century, and is now mainly grown and eaten there. The Latin name derives from that of the notorious Captain Bligh, who first brought it to the islands, as he did the BREAD-FRUIT. The fruits are about 3 inches (7 cm) long and grow on a large tree. The pink pods open, exposing black seeds encased in a fluffy yellow pulp which resembles scrambled egg. This pulp is the only part of the vegetable which is edible; the pink tissue is poisonous, as is underripe fruit which has not opened. For this reason, ackee is almost always sold canned outside its native habitat. If you find it raw, make sure you discard everything but the yellow pulp, and search the pulp well for traces of pink fibre.

The classic dish using this vegetable is codfish and ackee. Although the codfish used is traditionally salted, the recipe can also be made with fresh fish.

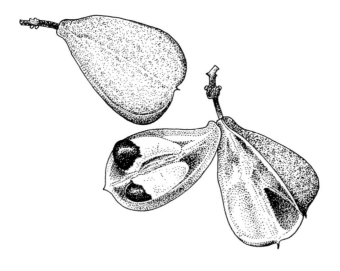

CODFISH AND ACKEE

1 lb fresh ackee pulp or canned ackees	500 g	1 lb
½ teaspoon salt	½ tsp	½ tsp
1 lb fresh cod fillets	500 g	1 lb
1 oz butter	25 g	2 tbsp
2 tablespoons oil	2 tbsp	2 tbsp
2 onions, sliced into rings		
1 large tomato, chopped		
2 slices chilli peppers or 2 scotch bonnet peppers, seeded		
1 sprig thyme		
2 hard-boiled eggs, quartered		
8 black olives, pitted		
½ teaspoon paprika	½ tsp	½ tsp

Rinse fresh ackees in cold water, or drain canned ackees. Pour 3½ pints (2 litres/2 quarts) ackees, if raw, and the fish. Boil, uncovered, until the fish is cooked and the ackees tender — about 20 minutes. If the ackees are canned, add them in the last few minutes of cooking the fish.

In a frying-pan (skillet), heat the butter and oil together. When hot add the onions, and cook until they are transparent. Add the tomato, pepper and thyme. Sauté for 5 minutes.

Drain the ackees and codfish. Flake the fish and add it to the contents of the frying-pan. Cook for 5 minutes. Transfer the ackees to a serving dish, pile the codfish mixture in the centre and garnish with hard-boiled eggs and olives. Sprinkle the eggs with paprika.

4 servings

Artichoke *Cynara scolymus*

Globe artichoke

The artichoke originates from the Mediterranean area and is a member of the thistle family. It consists of the unopened flower and usually part of the stem of the plant. The petals, usually referred to as 'leaves', are thick and green like leaves, and they become soft and edible at the base when cooked ('artichoke bottoms'). When all the leaves are pulled away, the choke is revealed. This is a hairy tuft in a mature artichoke, though it is soft enough to be edible in a baby artichoke. Beneath the choke is a fleshy pad, the heart, which is the most delicious part of the vegetable.

The Greeks and Romans ate artichokes, or members of the same family, so they have been popular for a long time. Although they grow best in warm climates, they have been known to grow as far north as Great Britain. They are very nutritious and reputed to be good for the liver. Artichokes are in season in late spring and early summer.

Choose artichokes that have flat, glossy leaves, and avoid those with shrivelled tips. Wash them well, or soak in a bowl of salted water to flush out any insects.

The easiest way to prepare artichokes is simply to cut off the stem as near to the flower as possible, and trim it. Sprinkle all cut parts with lemon juice to prevent blackening. The leaves can be trimmed back with a sharp knife so that the water penetrates to the bottoms easily to cook them. The stem can be eaten, when the fibrous outer part is peeled away. Artichokes should be cooked in plenty of boiling water, with lemon juice, until soft. Test for doneness by inserting a sharp knife through the heart from underneath.

Artichokes are eaten by hand, by pulling off the leaves and dipping them in mayonnaise, hollandaise sauce or plain yoghurt. When you come to the choke, pull it off — it should come away cleanly — and eat the heart.

To prepare artichokes for stuffing, use a very sharp knife to trim away as much of the top part of the leaf as possible and pull out the central leaves and the choke. The centre can then be filled with meat and baked in the oven. However, because they are so tough, the easiest way to stuff artichokes is to parboil them for about 10 minutes first. Drain them and leave them to cool before cutting into them.

Asparagus pea

Psophocarpus tetragonolobus

Goa bean
Winged pea

This climbing plant, with its unusual-shaped pods and their attached leaves, which make them look as if they are about to take flight, is fairly new to Great Britain, though it has been grown in the United States for some time. The pods are deeply ridged, and look a little like blunt-ended OKRA, except that green 'frills' are attached to the ridges, adding to the illusion of wings.

Although all parts of the plant are edible, the young pea pods are the most popular; they taste a little like asparagus. Always look for very young pods, and eat them as soon as possible after purchase, as they dry out quickly and become woody.

Asparagus peas do not overwinter in cold climates, and so are at their best in summer. They are best steamed or boiled and served with butter, or in a stir-fry mixture.

This plant is not to be confused with the YARD-LONG BEAN, which is sometimes called the asparagus bean. In fact, the asparagus pea is a member of the trefoil family and a native of southern Europe.

To cook asparagus peas, simmer them for 15 minutes in very little water and dot with butter, margarine or yoghurt before serving.

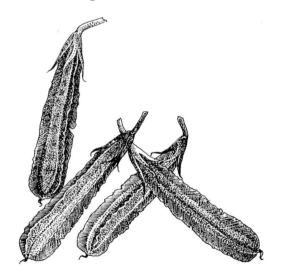

Bamboo shoots *Bambusa sp.*

Takenoko
Tung sun

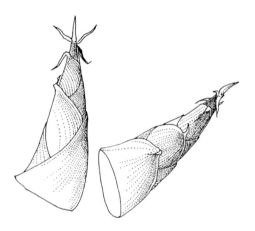

These shoots are an essential ingredient in Chinese cooking and can be bought fresh at Chinese shops, but are mostly only available canned. There are two types, spring and winter. The winter are smaller and tastier.

Bamboo shoots can be boiled if raw, then incorporated into stir-fried Chinese vegetable dishes. Use them like BEAN SPROUTS.

VEGETARIAN CHOP SUEY

1 small onion		
2 carrots		
8 oz bamboo shoots	250 g	½ lb
12 oz bean sprouts	350 g	¾ lb
1 tablespoon sunflower oil	1 tbsp	1 tbsp
1 garlic clove		
4 tablespoons soy sauce	4 tbsp	4 tbsp

Slice the onion very thinly. Cut the carrots and bamboo shoots into matchstick strips.

Heat the oil in a wok or heavy frying-pan (skillet). Crush the garlic into it, then add the onion and carrots. Stir-fry for a few minutes or until the onion has become transparent. Add the bamboo shoots and cook for 1 minute, then add the bean sprouts and stir-fry for another minute. Add soy sauce to taste. Serve immediately.

4 servings

Banana flower Musa sf.

This is the male flower of the banana plant, a fat, purple magnolia-shaped blossom with a fleshy centre, the tropical equivalent of an artichoke. It is eaten in southeast Asia, and in India as far north as Bangladesh, and can be found in oriental and Indian vegetable markets in Europe and the United States.

To cook the banana flower, put it in boiling water for 20 minutes, then discard the outer, purple leaves. The fleshy part can be eaten with mayonnaise like artichoke leaves, or can be boiled in thick, unsweetened coconut milk, then left to cool and eaten cold.

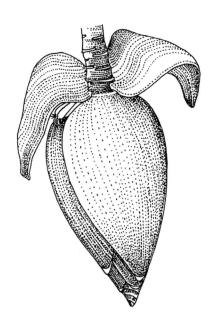

Bean sprout *Phaseolus aureus*

Black gram
Green gram (India)
Moyashi
Mung bean
Tu ya ts'ai

These sprouts are now sold commercially; in the past one had to sprout the mung beans oneself to obtain bean sprouts. The sprouts are an important ingredient in Chinese and Japanese cooking. In both cuisines, they are stir-fried rapidly in a wok in a sprinkling of oil to wilt them; they should still be crispy inside.

Bean sprouts will grow at any time of the year. Always choose firm, fresh ones, and eat them as soon as possible, or they will wilt or turn slimy. Bean sprouts are delicious raw in a salad.

The mung bean itself, a tiny green bean, can be boiled and eaten as a vegetable in curries. However, it requires at least 8 hours' soaking (discard the soaking water) and longer cooking than many larger beans, so you might feel that it is not really worth the effort.

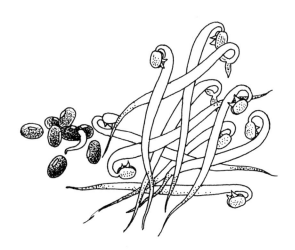

Bitter gourd *Momordica charantia*

Balsam pear
Bitter cucumber
Bitter melon
Karela

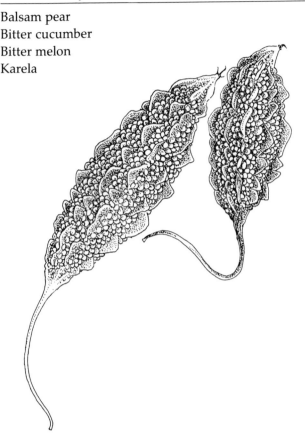

The bitter gourd is grown in India and wherever Indians have settled, and is a common ingredient in Indian dishes. It ripens to orange-yellow or white, but is usually sold at its green stage. It looks like a ridged, warty cucumber. Inside, there are reddish-brown seeds. The vegetable is peeled, sliced, the seeds discarded, and the slices soaked in salted water for an hour before cooking, to remove the bitterness. It is then used in curries, pickles and chutneys.

Choose small bitter gourds which are firm and not shrivelled, and free from blemishes or damage. They are in season in summer and autumn. Very closely related, and treated in the same way, is the **spiny bitter gourd** (*Momordica cochiniensis*). It is more rounded in shape and covered all over with pointed warts.

17

SPICY BITTER GOURD

4 small bitter gourds or spiny bitter gourds		
1 tablespoon oil	1 tbsp	1 tbsp
1 garlic clove, crushed		
4 curry leaves		
2 small hot green chilli peppers, chopped		
½ teaspoon ground turmeric	½ tsp	½ tsp
½ teaspoon chilli powder	½ tsp	½ tsp
1 teaspoon black mustard seed	1 tsp	1 tsp
1 teaspoon black onion seed	1 tsp	1 tsp
½ teaspoon salt	½ tsp	½ tsp
pinch of asafoetida (optional)	½ tsp	½ tsp

Slice the bitter gourd thinly and soak it in salted water for 1 hour. Drain and pat the slices dry. Heat the oil in a frying-pan or wok and add the garlic, curry leaves, and chillis. Cook, stirring constantly, for 1 minute. Add the bitter gourd and the rest of the ingredients and cook, stirring frequently, for 10 minutes or until the bitter gourd is tender. Serve with rice.

4 servings

Black-eyed peas *Vigna sinensis*

Black-eyed beans
Cow-peas

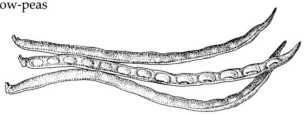

While black-eyed peas are a staple of the diet in the southern United States, these tropical pulses were unfamiliar in Europe until they were introduced by the West Indians. The peas are white, with a black eye and are bean-shaped, so that they are called peas or beans in different countries. As the Latin name implies, they were originally introduced to the western hemisphere from China.

BLACK-EYED PEAS AND RICE

1 lb black-eyed peas	500 g	1 lb
½ oz butter	15 g	1 tbsp
8 oz long-grained rice	250 g	½ lb
1 onion, finely chopped		
2 ripe tomatoes, skinned and chopped		
4 tablespoons chopped parsley	4 tbsp	4 tbsp
1 teaspoon salt	1 tsp	1 tsp
½ teaspoon black pepper	½ tsp	½ tsp

Put the peas in water to cover and soak for 2 hours. Drain, discarding the water. Put the peas into a pot and add fresh water to cover. Bring to the boil and simmer, partially covered, for 45 minutes to 1 hour or until tender.

After the peas have been cooking for 30 minutes, put the rice in a heavy pot with a tight-fitting lid and add 24 fl oz (750 ml/3 cups) water. Bring to the boil covered, then simmer on very low heat for 20 minutes.

Melt the butter in a frying-pan (skillet). Add the chopped onion and cook until transparent. Add the onions to the peas with the chopped tomato. Stir well.

Drain the rice and rinse it briefly under the cold water. Add it to the peas and stir well. Add the chopped parsley, salt and pepper and serve.

6 servings

Bok-choy *Brassica campestris* var. *sinensis*

Celery cabbage
Chinese cabbage
Mustard greens
Pak choy

These peppery leaves originate from southeastern China, but are popular in many sub-tropical climates, including the southern United States. The long thick stems bear wide green leaves at their tops. They are popular cut into pieces and fried, either stir-fried as in Chinese cooking or fried with meat as in southern American cooking.

Bok-choy is even more popular in Korea, where it is pickled with ground chilli pepper and a lot of garlic. The crispy fried Chinese 'seaweed', a Peking speciality, is often made with bok-choy.

Choose firm, undamaged leaves which do not look wilted. Separate the leaves from the stems before cooking, as the leaves will cook much more quickly. It can be bought throughout the year, especially in Chinese shops.

The name 'Chinese cabbage' is applied to both this vegetable and to the one better known as CHINESE LEAVES, but whereas the latter is eaten raw, the peppery taste of bok-choy tends to rule this out.

Breadfruit Artocarpus communis

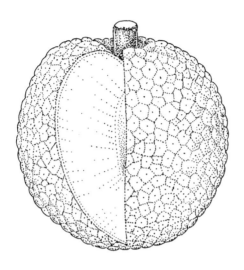

This fruit originates in the south Pacific, where it is generally eaten roasted. It is very large and heavy, and about 8 inches (20 cm) in diameter. The thick rind is divided into a mosaic-like pattern; it is yellow when young and ripens to green and brown.

The breadfruit has an important place in British history. It was Captain Bligh's concern for the breadfruit seedlings he was bringing to the West Indies from Tahiti on his ship, the *Bounty*, in 1787, which lead him to deny water to the crew rather than see the seedlings shrivel, thus causing the famous mutiny. The seedlings survived and breadfruit is today a popular vegetable in the West Indies, where it is usually boiled and eaten like YAM.

The best varieties of breadfruit are seedless, but if seeds are found they can be cooked and eaten. They grow inside the central core. The breadfruit consists almost entirely of starch and is a good substitute for bread. It can be sliced and oven-baked (wrapped in foil), baked in the embers of a barbecue, boiled, or sliced and fried. Breadfruit is very high in calories, with 105 to every 4 oz (100 g) portion. It is low in protein, fat and vitamin A, but is a good source of other vitamins and trace elements.

BREADFRUIT PUDDING

½ breadfruit		
¾ pint coconut milk, unsweetened	500 ml	2 cups
6 oz dark brown sugar	175 g	½ cup
1 teaspoon tamarind syrup (optional)	1 tsp	1 tsp
1 teaspoon cinnamon	1 tsp	1 tsp

Peel the breadfruit, remove the seeds if any, and cut into chunks. Add the chunks to a pot of salted boiling water. Reduce the heat, cover the pot and simmer for 20 minutes. Drain and leave to cool.

Butter a pie-dish. Preheat the oven to 350 °F (180 °C/Gas Mark 4). Mash the breadfruit by hand or in a food processor, and beat in the coconut milk, dark brown sugar, tamarind syrup (if used) and cinnamon.

Turn the mixture into the pie-dish and bake until the pudding is golden-brown, about 40 minutes.

4 servings

Burdock *Arctium lappa*

Gobo

This root grows wild and unnoticed in northern Europe and the United States, but it is a most popular vegetable in Japan and China and is gaining popularity among vegetarians and oriental food lovers in the west. The root is very long, a little like SALSIFY, and it can be cooked in the same way, or as a potherb with carrots, turnips and onions. The young, spade-shaped leaves can be eaten like spinach. The Japanese often pickle burdock and serve it as a suimono salad.

Both leaves and roots can be found in Chinese and Japanese vegetable markets.

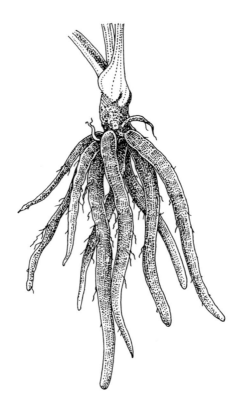

Calabrese Brassica oleracea var. *italica*

Green or purple-sprouting broccoli

The calabrese is a variety of broccoli—one which to non-experts is virtually indistinguishable from broccoli itself, but the tightly-packed green or purple flower buds of calabrese have a particularly delicate flavour. The name is Italian and simply means 'from Calabria', a town in Sicily. However, calabrese is now grown mainly in California for the American market, and in Jersey and Guernsey in the Channel Islands for the European market. Calabrese was introduced into France and Britain in the early eighteenth century from Italy, but was rarely eaten in the United States until after World War I.

Calabrese, like broccoli, has become popular partly because of its high vitamin content, so highly-prized during the current preoccupation with health, and is considered particularly beneficial to children in the way that spinach once was. Like all green vegetables, it has a high vitamin C content and many trace minerals, but it is also high in some of the B vitamins.

Before cooking, calabrese should be trimmed of the tiny leaves and side shoots and thoroughly washed. It should then be thrown into a large pot of boiling water and cooked just until the stalks are tender, about 10 minutes. Alternatively, it can be put in a deep dish containing a tablespoon of water, covered with cling film (plastic wrap) and microwaved for about 5 minutes.

Calabrese is delicious eaten hot with butter or cold in a mixed salad with mayonnaise. It must always be cooked, as the stems are too tough to eat raw.

Callaloo Amaranth sp.

Amaranth
Bhaji (India)
Elephant's ears
Sag
Sagaloo

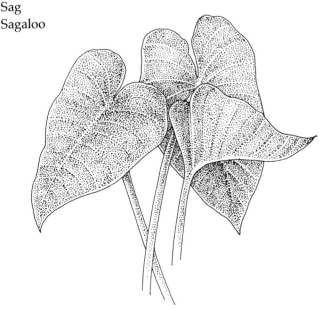

Several juicy leaves are sold under these names, and all are varieties of the species *Amaranth* which originates from Africa. **Surinam amaranth** or **Chinese spinach**, which probably originated in the Caribbean, is a tall plant with large, wide leaves. Like spinach, it is rich in iron and vitamin C. The DASHEEN leaf is also called callaloo and is prepared in the same way. For some inexplicable reason, all the members of the amaranth family, wherever they grow, from Greece to Mexico, have been attributed with magical properties and were used in religious rites. The Aztecs offered them as food to the gods together with the sacred maize plants. The ancient Greeks also revered the amaranth as having amazing curative powers.

All members of the family resemble spinach or sorrel, and all are used in cooked vegetable dishes. The most famous of these is a soup called Callaloo itself, which is a speciality of Trinidad and the French Antilles.

One can occasionally find fresh callaloo leaves imported from the West Indies, usually in summer. At other times they are only found canned. If you cannot get fresh ones, use sorrel or other similar types of leaf (see MELOUKHIA).

CALLALOO

8 oz callaloo leaves, well washed	250 g	½ lb
1 oz butter	25 g	2 tbsp
1 onion, finely chopped		
1 garlic clove, finely chopped		
1¼ pints chicken stock (broth)	750 ml	3 cups
3 fl oz unsweetened coconut milk	85 ml	6 tbsp
1 teaspoon salt	1 tsp	1 tsp
8 oz fresh white crabmeat	250 g	½ lb
Dash of Tabasco, pili-pili or Pickapeppa sauce*		

Roughly chop the callaloo. Melt the butter in a large frying-pan (skillet). Add the onions and garlic and cook until the onion is transparent. Add the callaloo and stir for 5 minutes, until they are evenly coated with butter.

Stir in the stock (broth), coconut milk, and salt. Bring to the boil, then cook over high heat, uncovered, for 10 minutes, to reduce the liquid. Add the crabmeat, and Tabasco, pili-pili or Pickapeppa sauce to taste. Stir to heat the crabmeat through, then serve immediately.

4 servings

*Jamaican seasoning available from supermarkets in Britain and the U.S.

Cardoon *Scolymus cardunculus*

This relative of the artichoke grows well in southern France and other parts of the Mediterranean and will even flourish in cooler climates, though it seems to have been very much neglected in the English-speaking world.

Unlike the artichoke, the cardoon's thistle-like flower is ignored, and only the stems are eaten. They resemble dry, silvery celery, with a rather downy stem, which is reputed to curdle milk. Cardoons are in season in late autumn, and, like artichokes, are prone to blackening when cut, so they should be prepared with plenty of acidulated water nearby in which to drop them.

Buy juicy-looking stems and avoid those which look too dry or stringy. They do not keep well, as they loose moisture. Cardoons need long, slow cooking, which is another reason why they may not be too popular among the fast-food-loving British and Americans.

Cardoons are a popular Christmas vegetable in Provence. Here is how they are prepared:

BAKED CARDOONS À LA PROVENÇALE

1 lb cardoons	500 g	1 lb
2 tablespoons vinegar or acetic acid	2 tbsp	2 tbsp
2 tablespoons oil	2 tbsp	2 tbsp
2 tablespoons flour	2 tbsp	2 tbsp
16 fl oz milk	500 ml	2 cups
½ teaspoon salt	½ tsp	½ tsp
½ teaspoon grated nutmeg	½ tsp	½ tsp
4 oz grated Cantal or Cheddar cheese	125 g	6 tbsp

Fill a bowl with water and add the acetic acid. Peel off the fibrous parts from the cardoons, rather as you would with rhubarb. Cut the cardoons into 2-inch (5 cm) slices, and drop them in the water. Put a large pot of water on to boil. Add the cardoons and boil for 30 minutes.

Heat the oil in a saucepan. When it is hot, stir in the flour and cook to a smooth paste for about 3 minutes. Add the milk all at once and stir and bring to the boil. Add the salt, and cook, stirring constantly, until the sauce thickens. Add the nutmeg.

Oil an ovenproof dish and arrange the cardoon slices neatly in the dish. Pour the sauce over them. Bake in a preheated 350 °F (180 °C/Gas Mark 4) oven for 20 minutes. Sprinkle with the cheese and return to the oven for 5 minutes or brown under a hot grill.

4–6 servings

Cassava _Manihot utilissima_

Manioc
Tapioca
Yuca
Yucca

This tropical tuber, a native of South America, is the most popular of all the tuberous roots which substitute for the potato in hot countries and, as its Latin name implies, it has been put to many uses. There are 'sweet' and 'bitter' varieties; the 'bitter' varieties contain large quantities of hydrocyanic acid and are extremely poisonous when eaten raw, but the poison can be leeched out by grating the root and boiling it in several changes of water.

The thick, hairy outer skin covers a dense, white flesh which consists almost entirely of starch with traces of calcium and other minerals. It can be bought fresh all year round, and it is also exported cut into convenient pieces and frozen. It is a staple in Brazil and Colombia, and many countries in West Africa. In southeast Asia, the cassava is called tapioca, even when it is not sold in flakes and made into puddings. In Brazil, a shiny white flour called _farinha de mandioca_ is made from cassava. It is eaten with stews, as a side-dish, and used as a thickener; a similar flour is made in Nigeria and called _gari_.

To cook cassava, peel it, cut it into chunks and boil it, then mash it, or slice it thinly and fry it. Serve it with other fried vegetables, as an appetizer, or incorporate it boiled into a stew.

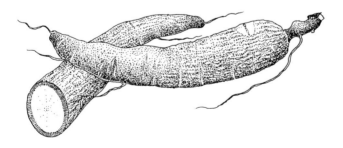

CASSAVA PUDDING

16 fl oz milk	500 ml	2 cups
2 tablespoons butter	2 tbsp	2 tbsp
10 oz sugar	300 g	1¼ cups
½ teaspoon salt	½ tsp	½ tsp
3 eggs, 2 of them separated		
4 tablespoons cassava starch (farinha or gari), mixed with 4 fl oz milk	4 tbsp 100 ml	4 tbsp ½ cup
½ teaspoon vanilla essence (extract)	½ tsp	½ tsp

Butter a pie-dish. Preheat the oven to 350 °F (180 °C/Gas Mark 4). Put the milk into a saucepan and bring it to the boil. Add the butter, sugar and salt, and stir well. Remove the pan from the heat and stir in the cassava mixture. Beat one of the eggs with the two yolks and stir this into the mixture.

Return the pan to the heat and stir continuously until the mixture thickens. Pour it into an oven-dish. Bake until the mixture is golden brown, about 30 minutes.

Whip the egg whites into stiff peaks and pile them on top of the mixture. Bake until the meringue starts to brown, about 15 minutes.

4 servings

Celeriac *Apium graveolens* var. *rapaceum*

Root celery

This tasty root deserves to be better known in English-speaking countries. It is a variety of celery which grows only spindly branches, and it has a large, round tuber root with numerous small projections on the surface. It originates from the Mediterranean and prefers a warmer climate to celery. The flesh is white and strongly-flavoured of celery. It is rich in calcium and phosphorus.

Celeriac is a winter vegetable which is particularly popular in Germany and central Europe. It is used as a potherb in stews (like the root of the parsley plant, which is generally discarded in English-speaking countries), but is particularly delicious when grated raw and served in salads. Here is a classic German recipe.

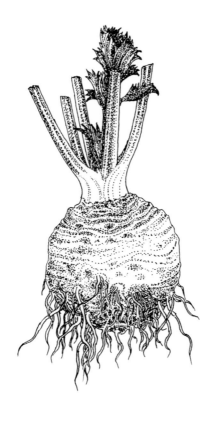

CELERIAC SALAD

1 lb celeriac	500 g	1 lb
2 large dessert apples		
juice of 1 lemon		
½ teaspoon salt	½ tsp	½ tsp
½ teaspoon sugar	½ tsp	½ tsp
5 fl oz sour cream	150 ml	⅔ cup
3 tablespoons slivered almonds	3 tbsp	3 tbsp
grated orange rind		

Peel the celeriac and grate it into a bowl. Grate the apples, peeled or unpeeled, and stir them in. Sprinkle with lemon juice, salt and sugar. Stir in the sour cream. Transfer to a salad bowl.

Toast the almonds in a dry frying-pan (skillet), watching them carefully to ensure they do not burn; when they start to give off their aroma, they are ready. Sprinkle them over the salad. Grate a little orange rind on a fine grater over the salad.

4 servings

Chayote *Sechium edule*

Brionne (France)
Cho-cho (Australia)
Christophine (West Indies)
Mango Squash
Mirliton (Creole)
Pepinello
Vegetable pear
Xoxo

Despite the numerous names for this pear-shaped member of the pumpkin family, originating from South America, it is not terribly popular outside the countries in which it is grown, though it is quite widely eaten in Australia, the West Indies and Louisiana. As with most of the squash family, to which it is related, the chayote is very easy to grow, the vines producing abundant crops in many types of soil and fruiting almost year-round.

The pear-shaped fruit is about 4 inches (10 cm) long; the outer skin is usually pale green, though paler and darker varieties are grown in tropical America. The flesh is cream-coloured and there is a single large seed in the centre. It is an excellent dieter's substitute for the avocado, as it can be subjected to the same type of treatment (though always after cooking), but it has virtually no calories. For those not on a diet, it can be thinly-sliced and deep-fried in batter. A soup is made from it in Nicaragua.

The chayote vine leaves are also edible, as is its fleshy root, which is low in calories and contains some vitamin C and some trace elements.

Buying guide: chayotes are furrowed and slightly pitted by nature, but should not look as though these indentations have been made by external forces, such as birds or tools. They should not look shrivelled, and should be completely firm to the touch. Look for them in good supermarkets and stores specializing in West Indian food.

STUFFED CHAYOTE

2 chayotes		
8 oz cooked chicken	250 g	½ lb
2 tablespoons soft breadcrumbs	2 tbsp	2 tbsp
¼ teaspoon ground cloves	¼ tsp	¼ tsp
1 teaspoon lemon juice	1 tsp	1 tsp
2 tablespoons melted butter or margarine	2 tsp	2 tsp
2 teaspoons grated Parmesan cheese	2 tsp	2 tsp

Parboil the chayotes in their skins in plenty of water for 20 minutes or until soft enough for a knife to penetrate them easily. Better still, microwave them on high for 5 minutes or more, until soft. Meanwhile, grind the chicken with the breadcrumbs, ground cloves and lemon juice.

Split the chayotes neatly in half and scoop out most of the flesh, leaving a ¼-inch to cover the shell. Chop the scooped-out flesh and mix it with the chicken mixture. Pile this into the shells, mounding it, and patting it down. Sprinkle with the melted butter or margarine and the cheese, and brown briefly under a hot grill (broiler). Serve hot.

Dieter's Tip: Instead of adding the butter and cheese, heat some tomato juice or puréed tomatoes and pour a little over the chayotes to make a sauce.

Left: chayotes, also known as cho-chos, showing the whole vegetables and the interior.

Chicory *Cichorium intybus*

Belgian endive (U.S.A.)
Endive (France, U.S.A.)

There is a lot of confusion about the naming of this plant. In the U.S.A. it is probably most familiar as endive or Belgian endive. It is indeed popular in Belgium. The French too call it *endive*, perhaps because what the British call 'endive' they call *chicorée*! The two plants are related (they are members of the Dandelion family), and the roots of both have been used as a coffee substitute, but there the resemblance ends.

This vegetable consists of the forced heads of the plant. Its pointed leaves are white with green or yellow tips, and have a characteristically bittersweet taste. They are sold in cans, but are infinitely better fresh.

Choose chicory that has a yellow tip; green indicates bitterness. Most books tell you to prepare chicory by cutting a cone-shaped piece from the base, as this part is supposedly very bitter. In fact, this is not the case, and merely trimming the base is sufficient.

Chicory is in season in January, February and March, when salad ingredients are hard to find, and it makes a delicious salad. It is also cooked by wrapping it in slices of ham, pouring a cheese sauce on top, then baking it.

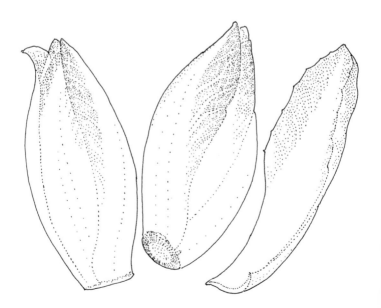

Chinese artichoke Stachys sieboldii

Chorogi
Crosnes

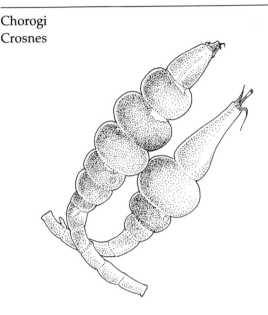

These little tubers resemble the JERUSALEM ARTICHOKE, hence the latter part of the name, but are not related to it. They are more regular in shape, and are the tubers of a Chinese plant. The name 'crosnes', which is often applied to these tubers in Europe, comes from the small French town of that name where they were first introduced in 1882, and from whence they were exported.

Chinese artichokes resemble Jerusalem artichokes in taste and can be used in the same way. Also like Jerusalem artichokes, they are hard to peel, but can also be eaten unpeeled. The simplest way to cook them is to boil them in salted water for 15 minutes, then serve them with butter and a sprinkling of fresh herbs. They are an interesting addition to a vegetable soup.

Chinese leaves Brassica pekinensis

Chinese cabbage
Nappa cabbage (U.S.A.)
Pak choy
Petsai
Shantung cabbage

The long pale-green leaves of this vegetable give it a resemblance to cos (romaine) lettuce, though it tastes like white cabbage. This Chinese vegetable was virtually unknown in Britain until the Israelis started growing it for export. The Californians got the same idea and now market it in the U.S. as Nappa cabbage, named after the Nappa valley, where it is grown.

Choose firm, white leaves, and eat them as soon after purchase as possible, so that they do not go brown and wilt. They can, however, be stored in the refrigerator for several days. Like all cabbages, Chinese leaves are rich in minerals and low in calories, so they are ideal for dieting. The leaves are so tightly packed that they can hold together when sliced crosswise into rounds. They can thus take the place of bread in an open 'sandwich', if meat and pickles are arranged on top of them, for a dieter's lunch or dinner. The vegetable can also be cut into shreds and stir-fried or boiled for a couple of minutes and tossed in butter.

Left: Chinese leaves (centre right), with the pak choi variant (centre, left) and the two kinds of chicory/endive: at the top is Cichorium endivia *and at the bottom* Cichorium intybus.

CHINESE LEAVES SOUP

2 tablespoons sunflower oil	2 tbsp	2 tbsp
1 green pepper, cored, seeded and diced		
2 onions, chopped		
2 celery stalks, chopped		
½ head Chinese leaves, chopped		
¾ pint chicken stock (broth)	500 ml	2 cups
1 tablespoon soy sauce	1 tbsp	1 tbsp
6 tablespoons plain yoghurt	6 tbsp	6 tbsp

In a deep pot, heat the oil and add the pepper, onions, celery and Chinese leaves. Cook on medium heat, stirring constantly, for 5 minutes. Add the stock and cook until the liquid comes to the boil. Add the soy sauce. Cover the pot and simmer for 15 minutes.

Remove half the vegetables with a skimmer and reserve them. Cook the rest for another 15 minutes, or until they are soft. Purée them with the liquid in a blender. Return the liquid to the pot and add the reserved vegetables. Reheat to just below boiling point. Swirl the yoghurt into the soup just before serving.

4 servings

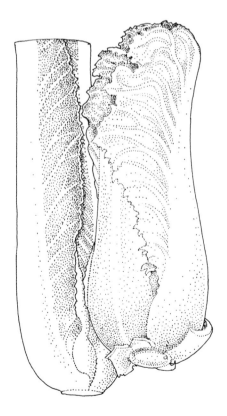

$\mathcal{D}$aikon *Raphanus sativus*

Mooli

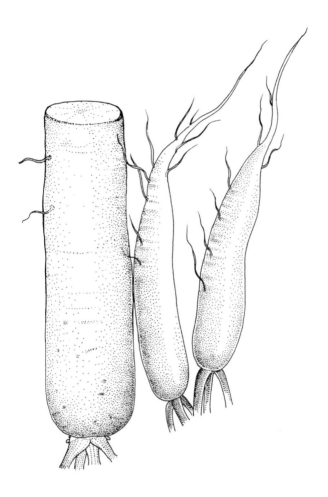

To those familiar with tiny red-skinned radishes, the long daikon, which looks like a white parsnip, hardly resembles a radish at all, but in fact it is merely another variety of the same plant. Outside of northern climates, radishes, even the round red ones, grow to a much larger size.

Daikon is the Japanese name for this type of radish, mooli the Indian and West Indian name. In Japan, it is grated and eaten as a condiment, with *sushi* (seaweed-wrapped rice containing fish and vegetables) and *sashimi* (raw fish) like the green horseradish called *wasabi*. In India, it is eaten cooked or raw.

Choose large, firm daikons, with fresh-looking green tops. To prepare daikon, scrape or peel off the skin and 'top and tail' it. Radishes are low in calories and contain a lot of vitamin C, as well as small amounts of iron and protein, so they are excellent food for dieters.

Another type of radish that is becoming well known is the **Spanish black radish**, which has the familiar shape but a black skin, which should be scraped off before eating.

A thin slice of daikon is a good substitute for mustard in a roast beef sandwich, and grated daikon is excellent in a salad. It combines well with fruit such as oranges and mandarins.

JAPANESE DAIKON SALAD

1 daikon, peeled and scraped		
1 carrot		
1 teaspoon salt	1 tsp	1 tsp
1-inch (2.5 cm) strip lemon rind		
2 tablespoons sugar	2 tbsp	2 tbsp
1/3 pint cider vinegar or rice vinegar	100 ml	1/2 cup
1 tablespoon lemon juice	1 tbsp	1 tbsp

Grate the daikon into a bowl, then grate the carrot. Sprinkle them with the salt and leave for 5 minutes. Rinse off the salt under cold running water, then squeeze and leave to drain.

Slice the lemon rind into matchstick strips. Mix the sugar, vinegar and lemon juice together. Add the lemon rind strips and mix well.

Squeeze the grated vegetables again and pour the marinade over them. Marinate for at least 1 hour, stirring frequently. Drain slightly before serving.

4 servings.

Right: a large daikon (white radish), with slices showing some decorative serving possibilities.

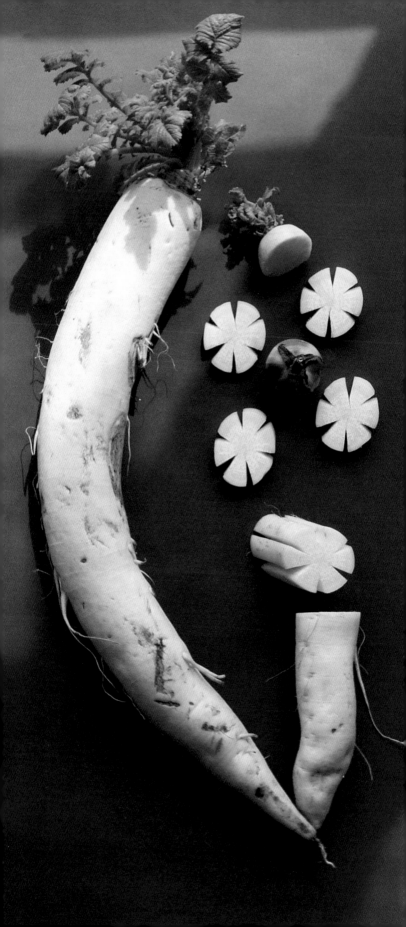

$\mathcal{D}$*asheen* *Colocassia antiquorum* var. *esculenta*

Taro

This round, hairy brown tuber has a skin which is divided into sections, as though it were articulated. The common Jamaican name for it, Dasheen, derives from *de Chine* ('from China'), as the root was imported from southeast Asia to feed the slaves in the West Indies. In Hawaii, where it was imported for the same purpose, it is generally known as taro.

The vegetable is prepared by scraping and peeling away the rough skin and cutting the flesh into slices. The flesh may be white, purplish or greenish. Occasionally, dasheen contains high concentrations of calcium oxalate, which may cause temporary discomfort when the root is eaten.

The dasheen is another member of the large family of tropical tuberous roots which take the place of the potato in countries where it is too hot to grow. It is eaten plain boiled with spicy sauces and stews. The leaves are also edible and are called CALLALOO.

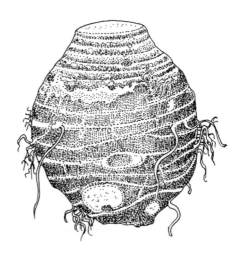

DOMINICAN FISH STEW

2 lb white fish fillets	1 kg	2 lb
1 teaspoon salt	1 tsp	1 tsp
1 lb dasheen	500 g	1 lb
2 spring (green) onions		
2 chayotes (q.v.)		
1/4 green papaya		
4 green bananas		
2 green plantains (q.v.)		
1 medium onion		
1 green bell pepper, left whole		
2 tablespoons oil	2 tbsp	2 tbsp
1 teaspoon chilli pepper	1 tsp	1 tsp
juice of 1 lime		
pepper		

Sprinkle the fish with salt and leave it while you peel and slice the vegetables. Heat the oil in a deep pot. Add the sliced onion and sauté until it is transparent. Add the rest of the vegetables and the fish, then add water to barely cover. Bring to the boil and simmer for 1 hour. Add the chilli pepper after 30 minutes, and the lime juice just before serving.

8 servings

ᗞoodhi *Lagenaria siceraria*

Bottle gourd
Doddy
Dudhi
Lokhi

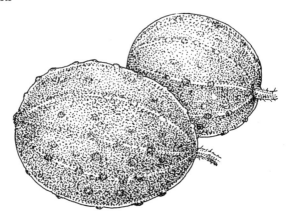

This vegetable is also called the bottle gourd, because when dried it is scooped out and used as a container. In India, however, young specimens are eaten as a vegetable. Since the gourd family is closely related to the pumpkins and squashes, gourds are cooked in the same way. The doodhi, which usually resembles a large pale-green cucumber (though occasionally rounded forms are found), has been cultivated for so long that its exact origins are unknown. Because it dries out so well, it can be preserved indefinitely, and remains of it have been found in Mexico dating from 7000 BC, and in Egypt dating from 4000 BC. It is still grown in both countries, and still used as a bottle.

Doodhi has a bland flavour and is a useful vegetable for adding to stews and curries. It is available in summer and autumn, mainly from Indian and West Indian shops. Choose vegetables which are free from blemishes and not withered.

To cook doodhi, peel it and remove any large seeds. All gourds can be cooked like the doodhi, including the **snake gourd** (*Lagenaria sp.*), which looks like the bottle

Right: several kinds of gourd, including a snake gourd (left), bottle gourd or doodhi (centre) and a loofah (right).

46

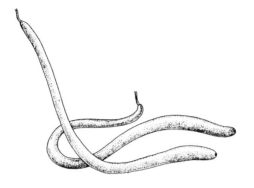

gourd but is longer and thinner. The **loofah** (*Luffa cylindrica*) is another member of the gourd family and, although fibrous when old, it is eaten when young in the same way as a gourd. It looks like a cucumber, but has deep, regular furrows on the surface. Loofahs are eaten mainly in China and the Caribbean. Although they are grown in the Mediterranean as decorative climbing plants, they are not eaten there, but are dried, peeled and made into the familiar abrasive bath sponges. Loofah can be prepared like doodhi and incorporated into Indian, Chinese or Caribbean dishes. The slightly bitter flavour can be removed by soaking in salted water for 1 hour before cooking.

SPICED DOODHI OR LOOFAH

2 doodhi or loofahs	approx. 750 g	1½ lb
2 tablespoons oil	2 tbsp	2 tbsp
1 teaspoon whole cumin seeds	1 tsp	1 tsp
2 curry leaves		
½ teaspoon salt	½ tsp	½ tsp
¼ teaspoon brown sugar	¼ tsp	¼ tsp
½ oz chopped fresh coriander (cilantro)	15 g	½ tsp

Peel and grate the doodhi or loofahs. Heat the oil in a frying-pan (skillet) or wok with a lid, then add the cumin seeds and curry leaves. Let them cook for a few seconds, then add the doodhi or loofahs and fry for 3–4 minutes, stirring constantly. Cover the pan and cook for 20 minutes, stirring occasionally to prevent sticking. Remove the lid. Add the salt and sugar and cook, stirring constantly, until all the liquid has dried up, and the mixture has begun to brown. Sprinkle with the fresh coriander before serving.

4 servings

Drumstick *Moringa oleifera*

The drumstick is an Indian bean with minute seeds, the fruit of a tree whose grated root, like others in the *Moringa* family, has a peppery flavour similar to horseradish. It grows in northern India, Pakistan and Bangladesh. The young shoots are succulent, but when purchased outside their native country tend to be rather tough on the outside. Nevertheless, they make an excellent basis for a vegetable soup.

To make drumstick soup, trim the drumsticks and cut them in half crosswise. Simmer for 20 minutes in water to cover. Reserve the cooking water. Split the drumsticks in half lengthwise, and scrape out the soft inner pulp. Return the pulp to the soup. Add some potherbs (onions, carrots, leeks, bayleaf) and simmer for another 20 minutes. Season to taste.

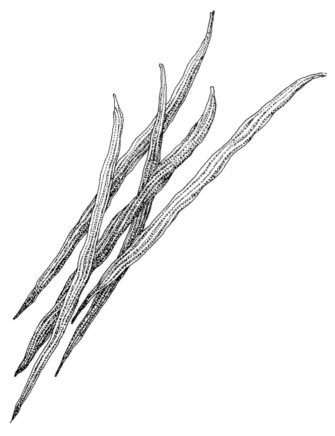

Eddoe *Colocassia esculenta* var. *antiquorum*

Malanga

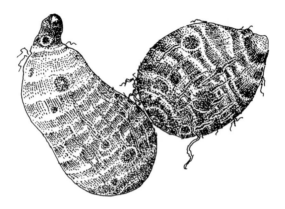

This is a roundish tuber grown in Africa and the West Indies, and very closely related to DASHEEN. However, some people prefer the taste to that of dasheen. The tubers of eddoes are smaller than those of dasheen. They can be cooked in just the same way. Eddoes are the classic ingredient of the West Indian stew known as pepperpot.

Another close relative of the eddoe is **Colocassa**, the version of the plant grown in Cyprus. It is probably closest to the type eaten by the Romans. Colocassa is a long root with a fairly shiny skin, marked into sections, and with a 'handle' at one end, which should be chopped off before cooking. It is best peeled and boiled whole. In Cyprus it is eaten stewed with chicken and lean pork.

Eggplant *Solanum melongena*

Apple of love
Aubergine (Great Britain)
Brinjal (Hindi)
Garden egg (Africa)
Guinea squash
Gully bean
Jew's apple
Pea aubergine
Susumber

In Europe and North America, the eggplant variety most often grown is violet to deep purple in colour and either the shape of a fat pear or of a long (up to 12 inches/ 30 cm) squash. The long, sausage-shaped purple variety is often known in the United States as Japanese eggplant. The eggplant actually comes in a vast variety of shapes and sizes. The reason for the original name becomes clear

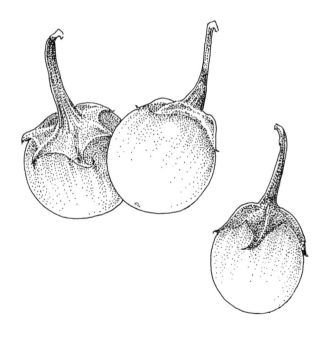

when one sees varieties from its native India which are pure white and exactly the shape of a hen's egg; these are sometimes delicately streaked with purple. In West Africa, 'garden eggs' are small, round and pale to deep yellow; in the Seychelles and the West Indies a closely related variety, whose Latin name is *solanum torvum*, has a skin that is pale green streaked with darker green, even when ripe. This type of eggplant is known in the West Indies as susumber, pea aubergine or gully bean.

Despite this enormous variety in shape and colour, eggplants can always be recognized by the tough spiny stalk and sepals, and by the thin skin covering a greyish-white flesh in which numerous tiny seeds are embedded. The vegetable grows on a thorny bush in any country or climate where the winters are mild. Consequently, the eggplant is eaten in a vast number of regions.

The eggplant is not particularly nutritious, and is low in calories. Choose specimens that are shiny and not shrivelled, and make sure the skin is not broken.

The main reason for the eggplant's popularity is its enormous versatility. The flavour of the eggplant changes each time it is prepared in a different way, and everywhere that it is grown new dishes have been developed to cater for local tastes. Eggplants contain a slightly bitter principle in the flesh, and large varieties are usually sliced, with or without the skin, sprinkled with salt and left for half an hour or so. The salt leeches out the bitterness. The eggplant slices are then rinsed and dried. Small eggplants are left whole or cut in half, and the salting is omitted. In some countries the eggplant is peeled, in others the skin is left on, and in the Middle East the eggplant is held over a naked flame to burn off the skin. This gives the flesh a smoky flavour when cooked.

In Turkey, the eggplant is stuffed for the famous dish called *Imam bayildi* (the Imam fainted); in Greece it is made into the national dish called *moussaka*, covered with cheese, noodles and ground meat. In central Asia, Romania and Israel it is turned into another dish of Turkish origin, *Baba ganoushe*, also known as Poor Man's Caviar. All over the Middle East, eggplant is salted, rinsed, drained, floured and fried in oil. In the Caribbean the small 'garden eggs' are sometimes stewed in coconut milk and sweet spices, then served with chicken dishes. In Japan and the Middle East, eggplants are pickled.

Eggplant recipes are legion. Here is one from the land where the eggplant originated.

SPICY INDIAN EGGPLANT

4 long (Japanese) eggplants or 6 'garden eggs'		
salt		
1 tablespoon white poppyseeds*	1 tbsp	1 tbsp
1 tablespoon cumin seed*	1 tbsp	1 tbsp
1 tablespoon fenugreek seed*	1 tbsp	1 tbsp
1 tablespoon coriander seed*	1 tbsp	1 tbsp
1 tablespoon black onion seed*	1 tbsp	1 tbsp
1 teaspoon mango powder (amchoor) or lemon juice	1 tsp	1 tsp
1 teaspoon chilli pepper	1 tsp	1 tsp
1 tomato		
1 small onion, chopped		
margarine or frying butter (ghee)		

Keep the eggplants whole, and neatly trim away the green stem. Split them neatly down the middle and sprinkle the cut halves with salt. Leave for 30 minutes.

Put the poppyseeds, cumin seed, fenugreek seed, coriander seed and black onion seed into a dry frying-pan (skillet). Cook until they begin to give off their aromas, stirring constantly. Grind the mixture to a powder in a coffee or spice grinder. Add the mango powder and chilli pepper to the mixture.

Drop the tomato in boiling water, remove and skin it. Cut it into quarters.

Add it to the spice mixture with the onion and mix well. Add 1 tablespoon melted margarine or frying butter (ghee), so that the mixture has the consistency of a lumpy paste.

Spread the paste thickly over the eggplants; there should be some left over. Heat a tablespoon of the margarine or frying butter (ghee) in a shallow pan with a lid. Add the eggplants, cut side up. Add the rest of the mixture to the pan. Sauté for 5 minutes, then turn the eggplants over and sauté another 5 minutes. Cover the pan and reduce the heat. Simmer for 20 minutes, removing the lid and stirring every 5 minutes to make sure the eggplants do not stick to the pan. Serve with plain boiled rice.

4 servings

*If any of these spices are unavailable they can be omitted. All are regularly stocked by Indian grocers.

Endive   *Cichorium endiva*

Chicory (France)
Curly endive
Escarole (U.S.A.)

This leafy plant, the shape of a rounded lettuce but with curly, narrow, ragged leaves, is a bittersweet salad vegetable. It is native to southeast Asia, but was introduced to Europe by the Dutch in the sixteenth century, since the Dutch were at that time the most partial to salads of any Europeans. Endive is best mixed with other salad leaves if eaten raw, as the flavour is rather bitter. Alternatively, it can be blanched (cooked very briefly in boiling water) to leech out the bitterness. Endive is a summer vegetable.

There is a variety of winter endive with larger flatter leaves called **Batavia** or **Batavian endive**, and **plain-leaved escarole** in the United States, after the part of Holland where it was developed. It is eaten as a salad green, like endive.

P.S. If you were expecting to find white, cone-shaped vegetables under this heading, you will find *that* kind of endive under CHICORY.

Fennel *Foeniculum vulgare*

Anise
Finnochio
Florence fennel

These large bulbs composed of intertwined thick-ribbed stems are the base of a large tall plant with feathery leaves like those of dill. Fennel seed (aniseed) is a herb, but the stems and bulbs still retain the strong flavour of aniseed that is so characteristic of fennel.

As its name implies, fennel comes from Italy and is grown throughout the Mediterranean. It is traditionally combined with fish in cooked dishes, but makes an excellent salad ingredient. Make sure the bulbs are not damaged, bruised or withered. The greener the plant, the stronger the taste, though some people find it a little overpowering in the young plants.

Fennel is in season in mid-summer. It is delicious when chopped in a salad and mixed with grated DAIKON and segments of mandarin orange.

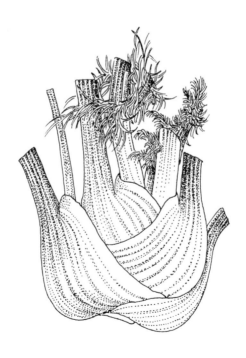

FENNEL WALDORF SALAD

2 *heads fennel*		
4 *small eating apples, quartered, cored and thinly sliced*		
2 *oz walnuts, coarsely chopped*	50 g	¼ cup
2 *tablespoons mayonnaise*	2 tbsp	2 tbsp
2 *tablespoons natural yoghurt*	2 tbsp	2 tbsp
1 *tablespoon lemon juice*	1 tbsp	1 tbsp

Trim the green leaves from the fennel and reserve them for garnish. Slice the fennel very thinly and put it in a large bowl. Add the apples and walnuts and stir well.

Beat the mayonnaise and natural yoghurt with the lemon juice. Pour this dressing over the salad and stir well until all ingredients are coated. Garnish with reserved fennel leaves.

6 servings

Fiddlehead fern *Matteuccia struthiopteris*

Ostrich fern

Many botanists claim that all ferns contain toxic compounds, though parts of varieties of fern are eaten. Bracken, for instance—the most common British fern— is said to have an edible rhizome (fleshy root).

The only fern eaten as a delicacy is the fiddlehead fern, which grows along the northeastern seaboard of the United States and Canada, especially in New Brunswick and Vermont. The fiddlehead is the tightly-curled frond of the ostrich fern. It grows in moist places such as river banks, and is only in season in late May. As soon as the fern begins to open and acquires a purplish tinge, it becomes toxic and cannot be eaten.

The fiddlehead fern has now become a delicacy eaten outside its native territory, for which purpose it is frozen or canned. But of course the preserved variety cannot rival the freshly-picked shoots.

Fresh fiddlehead ferns can be eaten raw in a salad or cooked just in their own washing water, like spinach, to soften them, then served hot or cold with butter or salad dressing. They are also a delicious addition to a vegetable soup.

Hamburg parsley

Petroselinum crispum var. *tuberosum*

Padrushka
Parsley root
Turnip-rooted parsley

This type of parsley is cultivated for its root, which is knobbly and whitish like a small CELERIAC. Hamburg parsley has never been widely grown in English-speaking countries, but it is very popular in central Europe, where it is a standard potherb in every stew, like carrots and onions. Its strong taste has been compared to that of celery.

Hamburg parsley should be washed and scraped, then put whole into the stew-pot. If you find the flavour of the vegetable itself too strong, discard it before serving the stew.

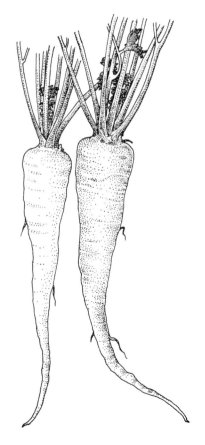

Heart of palm *Palmaceae sp.*

Palmito

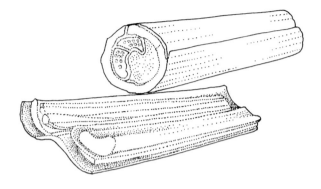

The hearts or young shoots are cut from several species of palm tree to produce this succulent vegetable. Unfortunately, it also completely destroys the tree. Hearts of palm are an important vegetable in Brazil, but have spread northwards to Florida and are now becoming popular in the fashionable cuisines of Europe and the United States. They have long been imported into France.

It is rare to be able to eat fresh hearts of palm outside Brazil, but the canned, marinated varieties are also excellent in salads. They consist of various layers of skin, like leeks, but have the delicate flavour of asparagus.

BRAZILIAN HEART OF PALM SALAD

3 ripe hard-shelled avocados		
6 hearts of palm		
2 hard-boiled eggs		
2 tablespoons olive oil	2 tbsp	2 tbsp
2 tablespoons lime juice	2 tbsp	2 tbsp
½ teaspoon salt	½ tsp	½ tsp
½ teaspoon freshly ground black pepper	½ tsp	½ tsp
2 oz chopped coriander	50 g	¼ cup

Carefully slice the avocados in half and scoop out the flesh, taking care not to pierce the shells. Chop the flesh roughly.

Slice the hearts of palm thinly crosswise and combine with the avocado. Chop the hard-boiled eggs and mix them in. Sprinkle the mixture with the olive oil, lime juice, salt and pepper. Pile it into the avocado shells and sprinkle generously with chopped coriander. Serve immediately.

NOTE For a more authentic Brazilian flavour, use palm oil instead of olive oil. It can be bought in shops specializing in African foods, and is a distinctive deep-orange colour.

Jerusalem artichoke

Helianthus tuberosus

The Jerusalem artichoke has nothing to do with Jerusalem and is not an artichoke! This knobbly little brown tuber is a native of North America, and is related to the sunflower. It is called 'artichoke' because the flavour is similar to that of an artichoke and 'Jerusalem' because this is a corruption of the Italian word for sunflower, *girasole*.

Choose fat, unwrinkled, unblemished tubers that are undamaged. Jerusalem artichokes are in season in late autumn and winter.

To prepare Jerusalem artichokes, scrub them well, then scrape and peel them and drop them as they are finished in water to which a tablespoon of lemon juice or vinegar has been added. They can be thinly sliced and fried, but are more often boiled for about 30 minutes. They are rich in minerals and are, like artichokes, reputed to be good for the liver. Jerusalem artichokes make a delicious soup, which the Victorians named Palestine Soup.

PALESTINE SOUP

2½ lb Jerusalem artichokes	1 kg	2½lb
2 onions		
1 stick celery		
1 oz butter	25 g	2 tbsp
1¾ pints chicken stock (broth)	1 l	1 quart
1¾ pints milk	1 l	1 quart
1 teaspoon salt	1 tsp	1 tsp
½ teaspoon white pepper	½ tsp	½ tsp

Slice the artichokes, onions and celery. Heat the butter in a deep pot with a lid and fry the artichokes, onions and celery until the onion is transparent, but do not let the vegetables brown. Add the stock, reduce the heat and cover the pot. Simmer until the vegetables are tender, about 30 minutes.

Strain the soup, and purée the vegetables in a liquidizer or food processor. Return them to the pan, and add the milk and seasoning. Bring to just below boiling point, and serve at once.

8 servings

Jicama *Pachyrrhizus erosus*

Yam bean

The European name, Yam bean, for this Central and South American vegetable is misleading, because it is neither a yam nor a bean. Jicama is an extremely popular vegetable in Mexico and the western U.S.A., particularly California, where it is generally eaten raw in salads, grated or whole. When raw, it tastes especially good sprinkled with lime juice and chilli powder.

Jicama is a large (as much as 10 inches/25 cm across) top-shaped tuber with a rough, brown inedible skin which peels away easily, even when raw, to reveal the juicy white flesh underneath.

Always choose jicamas which look firm and un-blemished and do not peel them until you are ready to eat them. Eat the whole vegetable at one time; when cut they tend to dry out and become hard and fibrous. They will keep for about two weeks in the refrigerator.

Jicama can be baked with a meat dish, like a potato, and even grated and used as a pie filling.

$\mathcal{K}ohl\mathit{rabi}$ _Brassica oleracea_

Rutabaga (U.S.A.)

Many people in the English-speaking world are put off this vegetable by its strange name. It also looks rather weird. It is a tuber, like the turnip, but the tuber shows above ground during growth. The vegetable sold in the shops has strange protuberances from it, which are in fact the ends of the stems where the leaves have been cut off. The outside skin is usually pale green, but is red in some varieties, and in Germany, where this vegetable is especially popular, a bright-blue variety called **azur-star** has been developed. The skin seems to be wrapped around the flesh in folds.

Kohlrabi is in season in the autumn. Choose glossy-looking specimens; they should be firm to the touch. To prepare kohlrabi, cut away the outer skin and cook the flesh, which has a similar consistency to that of a large radish (see DAIKON), but a milder flavour.

Kohlrabi is eaten mainly in the German-speaking countries. Here is a German recipe for it.

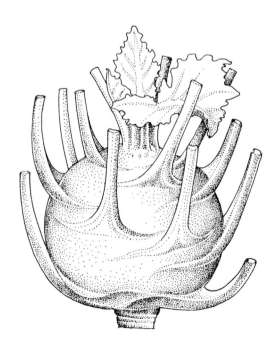

KOHLRABI WITH GERMAN SAUSAGE

2½ lb kohlrabi	1 kg	2½ lb
1 lb potatoes	500 g	1 lb
4 German or Polish boiling sausages		
½ oz butter	15 g	1 tbsp
½ teaspoon salt	½ tsp	½ tsp
½ teaspoon black pepper	½ tsp	½ tsp
4 tablespoons thick natural yoghurt	4 tbsp	4 tbsp
1 tablespoon flour	1 tbsp	1 tbsp
pinch of sugar		
4 tablespoons chopped parsley	4 tbsp	4 tbsp

Peel the kohlrabi and the potatoes and cut them into matchstick strips 1 inch (2.5 cm) long. Prick the sausages thoroughly with a fork. Heat the butter in a shallow pot with a lid. Add the vegetables and sausages and sauté briefly until the vegetables are lightly browned. Add water to cover, and the salt and pepper. Cover the pot and simmer for 20 minutes.

Mix the yoghurt with the flour and sugar and stir the mixture into the pot. Sprinkle with chopped parsley before serving.

4 servings

Lotus *Nelubium nuciferum*

The true lotus is a water plant related to the water-lily whose buds, leaves, seeds and roots are used extensively in Indian and Chinese cooking. All varieties of lotus, including the Egyptian lotus (*Nymphaea lotus*) are edible, as are water-lilies.

Lotus leaves are used as a wrapping for food in China and give it a unique fragrance. If bought dried, they should be soaked before wrapping around pieces of meat or fish, then steamed.

The tuberous roots (which resemble JERUSALEM ARTICHOKES or CHINESE ARTICHOKES) have an attractive lacy pattern when cut, which is used to full advantage in Japanese cooking. In India, the roots are boiled and mashed and used in *koftas* (vegetable rissoles) and curries. The Chinese and Japanese love them for their crunchy texture.

Lotus root is easy to find canned in Chinese and Japanese grocery stores, but can sometimes be bought fresh. It should be peeled, sliced and immediately dropped into acidulated water, as the flesh goes brown when it is exposed to the air. It makes an unusual vegetable similar in texture and flavour to WATER CHESTNUTS, and when mixed with soy sauce becomes a delicious savoury snack or Japanese-style salad.

Lotus seeds (also called nuts) are black. They should be picked unripe for preference, but if ripe they should be fried, boiled or roasted and the bitter germ removed. Roasted, they can be eaten as snacks or skinned, ground to a paste and sweetened with sugar to make a dessert.

Right: dried slices of lotus root (centre) with some dried lotus seeds around them.

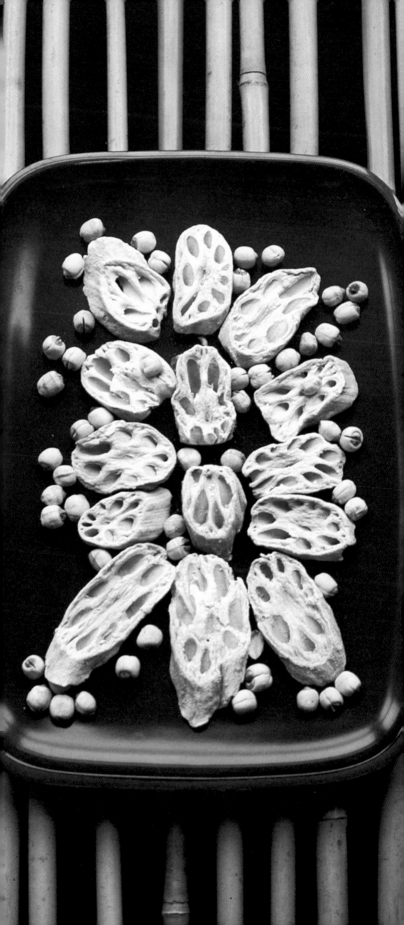

Mangetout _Pisum sativum_ var. _macrocarpum_

Snow peas
Sugar peas

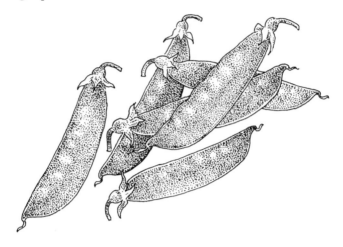

Mangetout are simply a variety of pea with a skin so thin and delicate that when it is young the whole pod is eaten, with the tiny peas inside it. Snow peas have to be 'topped and tailed', like beans, before cooking. They need vary little cooking, and should be slightly crunchy when eaten.

Mangetout are an important ingredient in Chinese cooking, and make a delicious cooked vegetable with roasts and stews. They should be stir-fried in a wok or frying-pan (skillet) in a little oil or butter for about 3 minutes, stirring constantly to coat them evenly with fat. Alternatively, they can be steamed in a steamer for 10 minutes.

Mangetout are rich in vitamin C and iron, but have only half the calories of garden peas.

Meloukhia *Corchorus olitorius*

Spanish okra

This leaf is related to the okra and the hybiscus, and is eaten primarily in Egypt and the Middle East, including Cyprus. Like okra, it has the effect of thickening the liquid in which it is cooked. Fresh meloukhia can sometimes be bought where there is a large emigrant community from its native habitat. It is in season from January to May when the land has been well watered with the winter rains. Otherwise, it is only available dried or canned.

Fresh meloukhia must be stripped from the stems on which it grows and well washed in several changes of running water, but do not wash it until you are just about to cook it.

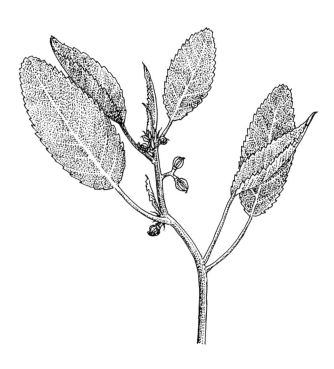

MELOUKHIA SOUP

2½ lb fresh meloukhia leaves	1 kg	½ cup/
or 4 oz dried leaves	125 g	2½ lb
3½ pints meat stock (broth)	2 l	2 quarts
½ teaspoon salt	½ tsp	½ tsp
1 oz butter	25 g	2 tsp
3 garlic cloves, crushed		
1 teaspoon ground coriander	½ tsp	½ tsp
½ teaspoon chilli pepper	½ tsp	½ tsp

Chop the leaves with a sharp knife and put them in a bowl of hot water. Leave them for 30 minutes or until they have swollen. Put the stock (broth) into a large saucepan and season it with the salt. Add the leaves; simmer them in the stock, with the saucepan partially covered, for 10 minutes if the leaves are fresh, 30 minutes if dry.

In a saucepan, melt the butter and add the chopped garlic. Cook, stirring, until the garlic is brown, then add the coriander and chilli pepper. Add this mixture to the soup, and serve immediately. Eat with pitta bread.

4–6 servings

Right: a stir-fry in a wok containing mangetout mixed with mung beans and bean sprouts, with peeled water chestnuts on top and bamboo shoots peeking through.

Nopales _Opuntia vulgaris_

Cactus leaf
Nopalitos

These are the leaves of the prickly pear cactus, whose fruits are also edible. Only the young leaves are used. The spines are removed (wearing gloves, to avoid touching the tiny, hairlike spines which surround the large spikes), although varieties without spines have been developed for ornamental and culinary purposes.

Nopales are an important ingredient in Mexican cookery and are sold all over the southwestern United States. When they come to market, they have been trimmed and chopped into tiny pieces. They can also be bought canned. Choose varieties that are not withered. They are cooked in stews or made into a salad, as in the following recipe, but raw nopales must always be cooked.

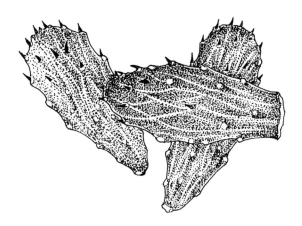

SALAD OF NOPALES

1 lb nopales, chopped	500 g	1 lb
1 teaspoon salt	1 tsp	1 tsp
2 tablespoons olive oil	2 tbsp	2 tbsp
1 tablespoon white wine vinegar	1 tbsp	1 tbsp
1/4 teaspoon black pepper	1/4 tsp	1/4 tsp
1/4 teaspoon chilli pepper	1/4 tsp	1/4 tsp
3 tomatoes, chopped		
1 onion, finely chopped		
1 tablespoon chopped fresh coriander (cilantro)	1 tbsp	1 tbsp

Wash the nopales thoroughly, and add them to boiling, salted water to cover. Cook them for 15 minutes, or until tender. Drain and discard the water.

To make the dressing, combine the salt, oil, vinegar and peppers. Combine the chopped tomatoes, onion and nopales. Toss the vegetables in the dressing and transfer to a salad bowl. Cover with clingfilm (plastic wrap), and chill for at least 2 hours. Sprinkle with the coriander before serving.

4 servings

Okra *Hibiscus esculentis*

Bindi
Gumbo
Ladies' fingers

These small, pointed, deeply-ridged green pods are full of seeds. The surface of the pods is sometimes slightly hairy. The plant originated in tropical Africa or India and was spread with the slave trade. The vegetables are picked at a very young stage; the pods can get much bigger, but they are then too tough for cooking. Okra is very popular in the southern United States and the Caribbean as well as in its native land. It is also grown in Thailand. The pods should be firm, unblemished and unshrivelled.

Although okra can be eaten raw, it is most often cooked. It has a rather glutinous texture when cooked, which adds bulk to soups and stews. In the Middle East, this glutinous texture is not appreciated, and the pods are soaked in lemon juice and salt, then fried to a crisp texture to get rid of any trace of sliminess. Do not cook okra in an iron pan or it will discolour, and do not overcook it or it will become unpleasantly slimy. Okra must be watched carefully when being boiled, as it is liable to make the water foam and boil over.

Here is an African recipe of the type which inspired the classic Louisiana dish, Okra Gumbo.

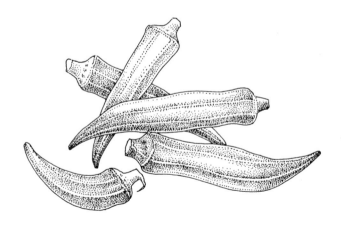

Right: okra, or ladies' fingers, shown whole and sliced.

74

AFRICAN CHICKEN GUMBO

1 lb okra	500 g	1 lb
3 tablespoons oil	3 tbsp	3 tbsp
1 chicken, cut into serving pieces		
4 tablespoons flour	4 tbsp	4 tbsp
2 onions, chopped		
4 tomatoes, peeled and chopped		
1¾ pints chicken stock (broth)	1 l	1 quart
1 teaspoon salt	1 tsp	1 tsp
1 teaspoon chilli pepper	1 tsp	1 tsp

Top and tail the okra, wash them and cut them into small rounds about ½ inch (1 cm) thick. Heat the oil in a large pot with a lid. Roll the chicken pieces in the flour, shake off the excess and sauté the chicken in the oil until it is evenly browned. Add the onions, tomatoes and the broth. Season with salt and chilli pepper and cover the pot. Cook on low heat for 45 minutes or until the chicken is tender. Add the prepared okra and cover the pot again. Simmer for another 10 minutes. Serve with plain boiled PLANTAINS or rice.

6 servings

$\mathcal{P}$epper *Capsicum anuum*

Capsicum
Chilli

The capsicum pepper is a native of the warmer parts of the New World, and has no relation at all to the black and white pepper (*Piper nigrum*) used as a seasoning. There are more than 100 varieties of peppers, and they are now an important ingredient of all the cuisines which feature strongly-spiced food, such as Chinese, Hungarian, Japanese, Indian and Thai.

Ripe peppers vary in colour from pale yellow, through red, to purple and chocolate brown, though red is the predominant colour. All have a tough, thin skin with a layer of flesh and a hollow centre. The centre contains a fleshy core to which a multitude of seeds are attached. Peppers vary in shape from bell-shaped to chunky, and some taper to a point. In general, it is the hot varieties that taper to a point, although some of the pointed variety eaten in Mexico are not sharp. Peppers also vary in size, the general rule being that the smaller the pepper, the hotter it is. Tiny green chillis, such as **chile pequin** and **chile arbol**, are eaten in Mexico, Japan, Thailand and India. The more rounded varieties, whose colours are a delightful range of brown, red, orange and yellow, known as **Scotch bonnets**, are popular in the West Indies. The Mexicans tend to eat peppers in their green, unripe state.

Larger peppers range in flavour from mild and sweet (**paprika peppers**, used in Hungarian cooking both fresh and ground to a powder, and **bell peppers**), to burning hot. Examples of the latter from Mexico are **Serrano**

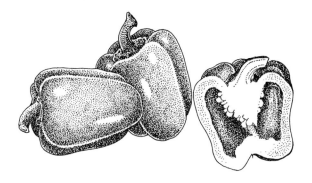

chillis and **New Mexico chillis**. Many peppers are eaten only in a dried state such as **chile cascabel**, a rounded variety whose seeds rattle inside (hence the name, which means 'rattle' in Spanish). Other hot peppers are the **chile serrano** and the **chile California** (also known as **Anaheim chilli**, which used to be grown most extensively on the site of what is now Disneyland in Anaheim, California). A popular chocolate-brown variety is the **chile poblano**. This last pepper is used to make the savoury sauce containing unsweetened chocolate, called *Mole Poblano*, which is Mexico's national dish. Hot peppers are the major ingredient of savoury spice mixtures and commercially-produced seasoning sauces such as Tabasco and piri-piri (which is the Portugese name for hot chilli peppers). They are ground and sold as a spice, known in most European countries and North America as cayenne pepper, though sometimes confusingly as chilli powder, which is often a blend of powdered chilli and other spices such as cumin seeds. In Egypt and the Middle East ground hot pepper is known as Sudanese pepper (*filfil soodani*).

Peppers have a high vitamin C content and are rich in other vitamins and minerals; the riper they are, the more nutritious. Many varieties are dried and used as sharp seasonings; the hot taste depends on the amount of a chemical called capsaicin in the pepper.

Capsaicin is a strong irritant. Thus it is important when preparing peppers, even mild ones, that you keep your hands away from your eyes. If you have a sensitive skin, it is advisable to wear gloves when handling peppers. Always wash your hands after handling peppers.

When buying fresh peppers, choose ones that are firm and glossy, and avoid any that are shrivelled. They are available all year round.

To prepare peppers, remove the stalk and slice down the centre, or cut round the stalk and pull it away with the core and seeds. The pepper can then be stuffed from below. The seeds should always be discarded when eating fresh peppers, as they are indigestible.

Peppers can be eaten raw in salads, or fried or stewed in vegetable mixtures or stews. They are often stuffed with rice, fish or meat mixtures. In the Mediterranean area, peppers are often charred before cooking (it is a

Right: sweet (capsicum) peppers of various shades, shown whole and sliced.

good idea to wrap them in damp newspaper first) to remove the thin skin.

This is a New Mexican recipe for stuffed green peppers.

CHILES RELLENOS

6 large green chilli peppers		
6 oz grated hard yellow cheese (Cheddar or Monterey Jack)	175 g	¾ cup
4 large tomatoes, skinned and chopped or 1 large can tomatoes		
2 garlic cloves		
½ onion, chopped		
½ teaspoon oregano	½ tsp	½tsp
3 tablespoons lard	3 tbsp	3 tbsp
8 fl oz chicken or beef stock (broth)	250 ml	1 cup
3 eggs, separated		
1 tablespoon flour	1 tbsp	1 tbsp
½ teaspoon salt	½ tsp	½ tsp
4 fl oz vegetable oil	125 ml	½ cup
1 tablespoon chopped fresh coriander (cilantro)	1 tbsp	1 tbsp
8 fl oz sour cream	250 ml	1 cup

Put the peppers under a hot grill (broiler) to sear the skins, then quickly put them into a brown-paper bag and leave for 10 minutes. The thin skins should then be easy to peel off. Slit each pepper lengthwise and remove the core and seeds, but not the stem. Stuff each with grated cheese.

To make the tomato sauce, skin the tomatoes if they are fresh and grind them with the garlic in a blender. Melt the lard in a frying-pan (skillet) and add the onion. Cook until transparent, then add the oregano and the puréed tomato. Cook, stirring constantly, for 5 minutes, then add the stock (broth). Simmer for 10 minutes and season to taste.

Make a batter by beating the egg yolks with the flour until thick; whip the whites separately with salt and fold them into the batter.

Heat the oil in a heavy frying-pan (skillet) or wok. Make sure each pepper is dry before dipping it in the batter by the stem, or the batter will not stick. Holding the peppers by the stems, drop them into the oil, in batches if necessary. Fry until lightly browned on both sides. Drain well on paper towels.

Pour the tomato sauce into a large serving dish. Arrange the peppers on top. Sprinkle with coriander (cilantro) and pass the sour cream separately.

6 servings

Pigeon pea Cajanus cajan

Cajan pea
Red gram (India)

These reddish peas grow in a twisted pod, and are usually sold dried or canned outside the tropics. In Africa, India and the West Indies the pigeon pea is an important article of diet.

Like all dried peas and beans, dry pigeon peas should be soaked for at least 2 hours in water to cover. Drain them and discard the water. The soaking not only softens the peas but also removes certain harmful chemicals present in legumes.

Here is an Indian recipe using pigeon peas.

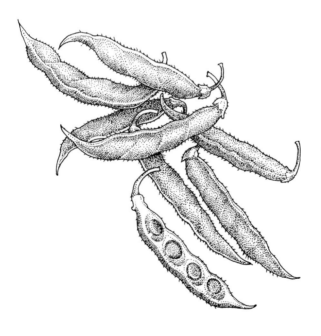

RED GRAM DAL

4 oz pigeon peas, soaked in water for 2 hours, drained	125 g	½ cup
1 pint water	600 ml	2¼ cups
½ teaspoon turmeric	½ tsp	½ tsp
1 green mango, peeled and sliced		
½ teaspoon salt	½ tsp	½ tsp
3 tablespoons margarine	3 tbsp	3 tbsp
1 teaspoon cumin seeds	1 tsp	1 tsp
½ teaspoon cayenne pepper	½ tsp	½ tsp
1 tablespoon chopped coriander	1 tbsp	1 tbsp

Put the peas into a deep pot and add the water. Bring to the boil. Add the turmeric, mango slices and add salt. Half-cover the pan and simmer for 20 minutes.

Heat the margarine in a frying-pan (skillet), and add the cumin seed. Cook, stirring, until it gives off its aroma. Add the cayenne pepper. Add this mixture to the peas and cook for another 10 minutes. Serve sprinkled with chopped coriander.

4 servings

$\mathcal{P}$lantain *Musa paradisaica*

Adam's fig

The plantain is a type of banana which is only eaten cooked. Like all members of the family it originates from Africa, but has been spread all over the tropics, because it is a starchy staple which, like the various types of yams, can take the place of bread and potatoes in hot climates.

The most common varieties of plantain look like a large, straight banana. They ripen from green, through yellow, to almost black, but other varieties ripen to pink and dark red.

Ripe plantains look like overripe bananas, but make sure the skin has not been broken. Unlike bananas, plantains can be stored in the refrigerator for up to three days. Keep them in a plastic bag. They are in season all year round.

Plantains are cooked in both the green and the ripe stage. They can be boiled or fried. When cooked, the sweet banana-like flavour emerges. They are sometimes sliced into thin rounds, fried and eaten like potato chips. They can also be cut into chunks and boiled alone, for an African 'mash', or cooked in stews. They are staple foods in South America, tropical Africa and the West Indies.

Plantains are eaten green or ripe. When green they are very firm and difficult to peel, but the skin will come off in sections if you first score it through lengthways and crossways. If you are making fried plantain chips, slice the vegetable right through and soak it in salt water for 30 minutes. Then push the rounds of plantain out of the skin.

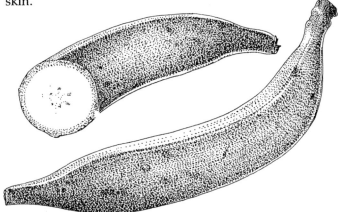

VENEZUELAN PLANTAIN CAKE

2 ripe plantains		
3 oz butter	75 g	⅓ cup
8 oz grated Cheddar or firm white cheese	250 g	1 cup
3 tablespoons sugar	3 tbsp	3 tbsp
1 teaspoon cinnamon	1 tsp	1 tsp
3 eggs, separated		
3 teaspoons dry breadcrumbs	3 tsp	3 tsp

Peel the plantains and slice them in half crosswise. Slice them lengthwise into thin strips. Melt half the butter in frying-pan (skillet). Add the plantains and cook, turning frequently, until the slices are golden brown on both sides. Transfer them to kitchen paper to drain.

Mix the cheese, sugar and cinnamon in a bowl. In another bowl, beat the egg yolks until they are thick and pale. In yet another bowl, whip the egg whites into stiff peaks.

Preheat the oven to 350 °F (180 °C/Gas Mark 4). Grease a ¾-pint (1 litre/1 quart) baking dish, and sprinkle it with the breadcrumbs. Fold the egg whites into the yolks. Ladle a quarter of this mixture into the dish. Cover it with a layer of fried plantains. Sprinkle with one quarter of the cheese mixture and dot it with pieces from the remaining butter. Repeat the layers of egg, plantains, cheese mixture and pieces of butter twice more until all the ingredients are used up, ending with egg mixture. Dot with the rest of the butter. Bake the cake in the middle of the oven for 40 minutes. Remove from the oven and cool slightly before turning out into a dish. Hot or cold, this dish is served with meat.

4 servings

Pumpkins and Squashes

Cucurbitaceae

Marrow (for squashes)

The *Cucurbitaceae* family includes not only pumpkins, vegetable marrows and squashes but also cucumbers, melons and CHAYOTES. The word 'squash', applied to all the marrow-type vegetables in this family, comes from an American-Indian word, and is used mainly in the United States, though as more American varieties are being imported and grown in Europe, the name is becoming commoner in Britain. The pumpkins and squashes produce their fruits on climbing vines. Most varieties flourish in hot climates, but a few can withstand the cool northern summers, and for this reason members of this family are found all over the world. Most varieties cultivated today originated in North America.

Pumpkins and squashes range in colour from dark green to pale greenish-white and deep golden-orange. They have skins which are often warty and furrowed and which should be discarded before eating. The raw flesh is too tough to be digestible so they should always be eaten cooked. The vegetables have a bland taste and need long

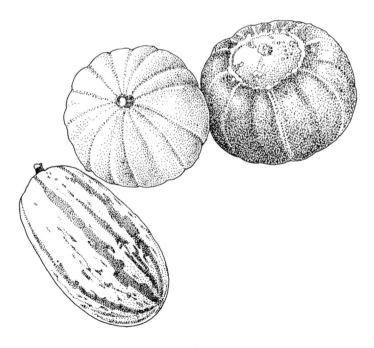

cooking times. Pumpkins have a hollow centre containing a spongy mass of fibres holding the seeds in place; in squashes and cucumbers the seeds are held in place within the flesh.

The pumpkin comes in a variety of colours ranging from pale yellow to deep orange, the latter being the colour of the American pumpkin (*Cucurbita mixta*). The American pumpkin takes longer to mature than the European varieties, so it is best grown in a warmer climate. European pumpkins (*Cucurbita pepo*) are generally pale yellow in colour and the flesh is less firm than the American variety; Russian pumpkins have white flesh and pale-green skins.

Pumpkins ripen in late autumn, which in northern America is convenient for Halloween (31 October). In Britain, the pumpkin lantern so popular in the U.S., has replaced the turnip lantern. Pumpkin pie is a traditional dish at Thanksgiving, the American festival which occurs in late November and celebrates the Pilgrim Fathers' first successful harvest.

The pumpkin is classified as a winter squash. Squashes are divided into two types: summer and winter squashes. In fact, both types are available throughout the year, but summer squashes are picked at a younger stage, and in general, the seeds are smaller and the whole vegetable, including seeds and skin, is edible. Summer varieties of squash include the **crookneck squash**, a yellow or orange squash with a neck bent into a hook; a small, flattened squash with a variety of names such as **custard squash** and **scalloped squash**; **custard marrow** or **pattypan squash**; and **table queen**, which varies in colour from

pale greenish-white to deep orange. The long green squash called a **vegetable marrow** in Great Britain is also a type of summer squash. So is the yellow vegetable whose flesh separates into spaghetti-like strands when cooked, which is called **vegetable spaghetti** in Britain and **spaghetti squash** in the United States. Miniature marrows are called zucchini in the U.S.A. and courgettes in Britain. They vary in colour from pale green to dark green flecked with light green.

Winter squashes (*Cucurbita maxima*) are tougher and more fibrous. They should be peeled before or after cooking, and larger specimens should be cut into chunks and the seeds removed before cooking. Varieties include the **butternut squash**, a bulbous, pale-yellow variety with orange flesh, and the **acorn** or **Des Moines squash**, a small green or yellow, rounded variety, as well as the **turban squash**, similar to the acorn squash, and the **Hubbard squash**, a very warty variety, eaten at the green unripe stage as well as in its orange, mature stage. New varieties of summer and winter squashes are constantly being developed, and are as popular for their attractiveness as a table decoration as for food.

Pumpkin seeds are very nutritious, being rich in fats and proteins. They are salted and sun-dried, and eaten as a snack in Mexico and the Middle East. They are also sold unsalted and skinned in health food shops in Europe and North America, where they are sometimes known by their Spanish name, *pepitas*. Use them like nuts. Try adding them to cooked vegetable dishes and to muesli.

Pumpkins and squashes are very versatile, because they can be eaten sweet or savoury, and their bland flavour brings out the taste of the other ingredients. They are rich in β-carotene and contain some vitamin C.

In addition to being made into the traditional Thanksgiving Pumpkin Pie, pumpkins and squashes can be incorporated into jams and chutneys, though their lack of natural acidity means that citric acid or some other souring ingredient should be added to ensure that the preserve does not deteriorate. They are also added to breads or cakes. In the Middle East, India and North Africa, pumpkin is eaten in meat stews. Pumpkins and squashes can also be puréed and added to soups, or boiled and served as a side-dish, with a white sauce.

Pumpkins and squashes should always be firm to the touch and be unblemished. They should not be shrivelled. Pieces of pumpkin can be bought, if the flesh looks firm and is protected from insects while on sale, but be careful to eat all members of this family only while they are fresh; they can deteriorate quickly. If they have an area on the skin that is paler than the rest, this is merely where they rested on straw or on the ground while growing and is not a blemish.

See also BITTER GOURD, DOODHI, SQUASH BLOSSOM.

Precooking pumpkins and squashes

Pumpkins and squashes need to be cooked until they are soft before they are added to other dishes. They have traditionally been cut into large chunks, peeled or unpeeled, and parboiled. However, this method leaves the vegetable soggy and waterlogged and removes most of the vitamins. The best way to precook a pumpkin or squash is to trim it, cut it in half, removing seeds and fibre—but not the skin—from the vegetable, then wrap it in foil and bake in a preheated 350 °F (180 °C/Gas Mark 4) oven for 30 minutes to 1 hour, depending on the size of the vegetable.

An alternative method, which is quicker and just as good, is to cut the vegetables into smaller chunks, put them in a dish and cover them with clingfilm (plastic wrap), then microwave them for 5 minutes. In both cases, all the flavour and as much of the nutritional value as possible are retained, and the skin can be peeled away much more easily than if it had been removed when the vegetable was raw.

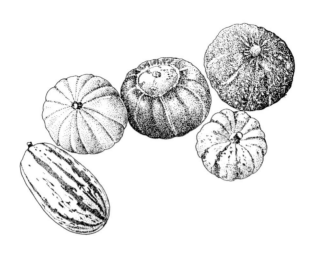

AFRICAN SUMMER SQUASH STEW

4 tablespoons oil	4 tbsp	4 tbsp
1 chicken, cut into serving pieces		
1 large onion, chopped		
2 large green bell peppers, seeded and chopped		
2 small hot red peppers, seeded and chopped		
4 oz unsalted peanuts, finely ground	125 g	½ cup
1 teaspoon salt	1 tsp	1 tsp
1 large or 2 small eggplant (aubergine or garden eggs), peeled and cubed		
1 summer squash (about 2½ lb/1 kg weight) or several small ones, peeled and cubed		

Heat the oil in a deep, heavy pot with a lid. Brown the chicken pieces and add the onions and peppers. Stir the peanuts into 8 fl oz (225 ml/1 cup) hot water, and stir this mixture into the pot. Add the salt, eggplant and squash. Cover the pot and cook at a simmer for 45 minutes, or until the chicken and squash are tender. Serve with rice.

6 servings

Radicchio *Cichorium intybus* var. *foliosum*

Red-leaved chicory

The radicchio is merely a red-leaved version of a type of endive popular in Italy, but has become so popular in yuppy circles (the vegetable equivalent of the Kiwi fruit?) that it is worth giving it a separate entry. The heads are tightly curled and wider than endive, so that it looks more like a form of lettuce. The flavour is identical to that of endive, i.e., bittersweet, and it is best included in a mixture of salad greens to allay the bitterness (and the cost!).

Radicchio originated in the Mediterranean and is imported from Italy, although it is also grown in California and other parts of the United States where there is a substantial Italian population. The heads of radicchio can also be baked or stir-fried.

RADICCHIO SALAD WITH BLUE CHEESE DRESSING

1 radicchio, leaves separated		
4 oz blue cheese	100 g	½ cup
1 tablespoon plain thick yoghurt	1 tbsp	1 tbsp
salt and pepper		
2 tablespoons sunflower seeds	2 tbsp	2 tbsp
1 teaspoon safflower or sunflower oil	1 tsp	1 tsp

Arrange the radicchio leaves on four salad plates. Mix the blue cheese and the yoghurt and season to taste. Stir in the sunflower seeds.

Divide this mixture between the four salad plates. Sprinkle lightly with the oil. Serve immediately.

4 servings

Rocket plant *Eruca vesicaria*

Arugula
Rokko

This Mediterranean plant is eaten in salads and as a herb. It is very popular in Italy and in the Italian émigré communities of the United States, as well as among Cypriots. It grows wild in colder climates, such as Britain, though it is not very popular there.

Rocket plant is rich in vitamin C and minerals. It makes a delicious addition to a salad with its strong, pungent flavour, slightly reminiscent of green coriander leaves. Choose leaves that are shiny and firm, and avoid any that are damaged or wilted. Rocket is available in the summer.

Salsify *Tragopon porrifolius*

Oyster plant
Vegetable oyster

This long, narrow root seems to have gone out of fashion, though it was highly prized in Victorian times. The delicate flavour of the flesh is said to be reminiscent of oysters, hence the alternative names. The root should be firm when bought, and must be cleaned carefully before cooking.

Salsify can be boiled, fried (cut into rings or strips), and even eaten grated raw. It makes a delicious raw salad when mixed with raw grated carrot and turnip or radish. Like its close relative, SCORZONERA it originates from Spain, but is now grown all over the northern hemisphere.

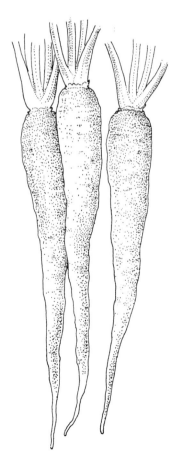

FRIED SALSIFY

1½ lb salsify roots	750 g	1½ lb
1 lemon		
½ teaspoon salt	½ tsp	½ tsp
2 oz butter	50 g	¼ cup
2 tablespoons chopped parsley	2 tbsp	2 tbsp

Scrape the salsify roots. Have ready a bowl of cold water into which you have squeezed half the juice of the lemon. Cut the roots into two or three pieces, crosswise, and place them in the water as they are peeled to prevent discoloration.

Bring a large pan of water to the boil and add the salt. Add the salsify and boil them until they are tender, testing with a fork after 30 minutes. Drain them and leave them to cool slightly. Slice them into rings about ½ inch (1 cm) thick and pat them dry.

Heat the butter in a frying-pan (skillet) or wok. Add the salsify rings and sauté them until they are lightly browned. Sprinkle with parsley before serving.

4 servings

Samphire Salicornia europea

Glasswort
Marsh samphire

This delicious vegetable, which grows on salt marshes and by the sea, is occasionally available in British markets. The plant is rich in sodium and was once used in the glassmaking process, hence the name. Samphire consists of long, asparagus-like branches, with several growing out at once from a tough brown stem.

Samphire should be prepared soon after cooking, though it will keep for a day or two in the refrigerator. Discard the stems and wash the green parts well in several changes of water.

Samphire is always eaten cooked. It is boiled just until done (do not overcook or it will lose its bright-green colour), and dabbed either with butter or eaten with hollandaise sauce, like asparagus, or pickled. The fleshy green part has a central woody core. To eat samphire, hold it in the hand and pull away the soft part from the woody core with the teeth.

Scorzonera   *Scorzonera Hispanica*

Black salsify

The name of this long black root vegetable means 'black serpent', quite an apt description. Nevertheless, this member of the daisy family has a delicate flavour. It can be eaten raw, but is best washed well and scraped and cut into rings, then cooked in lots of briskly-boiling salted water. The raw or sliced-and-cooked vegetable is also attractive in salads. It makes a delicate soup when simmered in milk with fresh herbs.

Scorzonera has much in common with its close relative SALSIFY. Both plants originate from southern Europe.

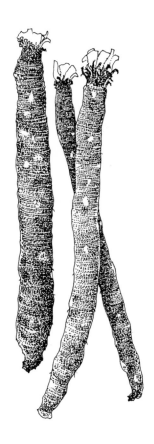

SCORZONERA IN BUTTER SAUCE

2½ lb scorzonera	1 kg	2½ lb
1½ oz flour	40 g	6 tbsp
2 tablespoons white wine vinegar	2 tbsp	2 tbsp
1¼ pints milk	750 ml	3 cups
½ teaspoon salt	½ tsp	½ tsp
1 medium onion, chopped		
1½ oz butter	40 g	3 tbsp
8 fl oz vegetable stock (broth)	250 ml	1 cup
½ teaspoon salt	½ tsp	½ tsp
1 tablespoon lemon juice	1 tbsp	1 tbsp
2 tablespoons double cream	2 tbsp	2 tbsp
1 egg yolk		
pinch of grated nutmeg		

Peel the scorzonera and cut them into 1-inch (2.5 cm) lengths. Have ready a bowl of cold water containing 1 tablespoon of the flour and the vinegar. Drop the salsify into the bowl as soon as they are ready, to prevent discoloration. Pour 8 fl oz (250 ml/1 cup) of the milk into a saucepan with about 1¾ pints (1 litre/1 quart) lightly salted water. Bring to the boil. Add the scorzonera and cook for 15 minutes. Drain (the liquid can be reserved for stock). Transfer the scorzonera to a shallow pan with a lid.

Melt the butter in a saucepan and sauté the onion until it is transparent. Add the rest of the flour, and stir to combine. Add the stock and milk and bring to the boil, stirring constantly. When the mixture boils, reduce the heat and cook, stirring, until the mixture thickens. Add the salt, nutmeg and lemon juice. Pour the sauce over the scorzonera. Cover the pan and simmer until the scorzonera are cooked through, about 40 minutes. Just before serving, beat the egg yolk and cream together. Remove the pan from the heat and stir them into the mixture. Transfer to a warm serving dish and sprinkle with grated nutmeg. Serve as a first course or a side dish.

4 servings

Seakale *Crambe maritima*

This summer vegetable grows wild along the seashores of northern and western Europe and bears some resemblance to broccoli. The stems are blanched and have a very delicate, almost asparagus-like flavour. Despite the name, it is no relation to **curly kale** (borekale), which is a member of the cabbage family and very hardy.

Seakale is not to be found much in shops, though it was marketed in Britain during and after the war, when other vegetables were scarce. Although the leaf tips can be eaten raw, the stems are the most prized part of the vegetable and need blanching (short cooking in salted water), and are then traditionally served in a napkin, with a sauceboat of melted butter for dipping them into.

$\mathcal{S}$eaweeds *Algae sp.*

There are no poisonous forms of seaweed: all are safe to eat, provided they grow in unpolluted waters, but as this is hard to guarantee nowadays it may be safer to buy seaweed that is grown commercially than to pick your own.

Seaweeds are eaten by many seafaring nations, but especially by the Japanese, Welsh and Irish. They are extremely nutritious in vitamins and minerals, while being low in calories.

Among the Japanese, **nori** and **kombu** (Japanese kelp, *Laminaria digitata*), known as **oarweed** in Britain, because it gets tangled up in oars, is used in the preparation of many Japanese dishes, but especially *sushi*, where the seaweed is used as a wrapping to hold the rice and vegetables firmly in place. In Wales, **laver** (*Porphyra umbilicus*) is fried and eaten for breakfast with bacon and eggs, particularly in the Swansea area. But first it needs prolonged boiling (for up to 4 hours) to soften it and break down the fibres. In Ireland, **carageen** (carragheen) or **Irish moss** (*Chondrus crispus*) is used to make soups and as a jelling agent for puddings, and is taken as a medicine as well as being eaten as a vegetable. In Jamaica, Irish moss, as it is called there too, is made into jellies and used as the basis of a punch-type drink of the same name.

Many seaweeds, especially carageen, are grown commercially for their jelling agents, known as alginates, of which one of the most important is lecithin. These are used in a wide range of processed foods and in the pharmaceutical industry. The gelatine called agar or agar-agar has long been enjoyed by vegetarians as a substitute for vegetable gelatine, and it has the advantage of jelling pineapple, which will not set in animal gelatine.

Dulse (*Palmaria palmata*) is a popular snack and is sold at fairs in Donegal and Ulster. It is first cooked and dried. It grows on rocks between the tide marks. **Sea lettuce** (*Ulva lactuca*) lives up to its name; pale green and very lettuce-like, it is common around the shores of Britain. It is eaten by some people as a vegetable.

In the following recipe, any other seaweed can be used instead of Japanese kelp.

JAPANESE CHICKEN SOUP

2 chicken breasts, skinned and boned		
4-inch (10 cm) piece dried Japanese kelp (kombu)		
2 tablespoons rice	2 tbsp	2 tbsp
4 Chinese leaves		
1 carrot, sliced thinly		
8 dried or fresh Japanese mushrooms (shiitake)		
8 oz tofu, sliced into 1-inch (2.5 cm) cubes	250 g	½ lb
2 leeks, sliced into 1-inch (2.5 cm) pieces		
7 oz vermicelli	200 g	½ lb

Cut the chicken breasts into bite-sized pieces. Wipe the Japanese kelp with a damp cloth to remove any sand, and cut it into 1½-inch (4 cm) squares with scissors. Put the rice into a muslin bag and tie it tightly.

Put the chicken breasts, kelp and rice into a deep pot and add 3½ pints (2 litres/2 quarts) water. Bring to the boil, then simmer for 30 minutes, skimming the surface to remove any scum. Remove the Japanese kelp, and add the rest of the vegetables; cook for 15 minutes. Add the vermicelli and cook for 5 minutes.

Pour into bowls and eat with chopsticks. Boiled rice can be served on the side.

8 servings

Sorrel Rumex acetosa

Sourgrass

Sorrel has a delicate texture and taste. The green, spade-shaped leaves resemble spinach, but grow in more northerly climates. Sorrel is very high in vitamin C. It withers quickly, so it should be cooked as soon as possible after it is picked or bought. The leaves can be sautéed in butter; they will 'melt' into a purée. The following is a classic Russian recipe for a chilled summer soup.

SHCHAV

1 lb sorrel	500 g	1 lb
1¾ pints water	1 l	1 quart
1 small onion, chopped		
½ teaspoon salt	½ tsp	½ tsp
2 tablespoons sugar	2 tbsp	2 tbsp
2 lemons, juice squeezed		
2 tablespoons sour cream or thick yoghurt	2 tbsp	2 tbsp

Wash the sorrel leaves well, rinsing them in several changes of water. Drain and chop them finely. Put them into a deep pot and add the water and salt. Bring to the boil. Add the onion and simmer for 15 minutes. Add the sugar and lemon juice and simmer for another 5 minutes.

Remove the pot from the heat and cool to room temperature. Pour into a bowl and chill. Swirl the cream into the bowl just before serving.

4 servings

$\mathcal{S}quash\ blossom$ Cucurbita Pepo

Courgette flowers
Zucchini flowers

The flowers of all the members of pumpkin, marrow and squash family are eaten wherever they are grown, usually as a main course stuffed with meat. The flowers are always yellow and are rich in vitamin C. Of course, it is rare to find the blossoms sold in the market, but home-grown squash produces a plentiful supply.

The squash family produces male and female flowers; the male flowers are bigger. The blossoms should be picked for cooking just before they open. They are delicious stuffed with ground meat, flavoured with raisins and pine nuts (*piñons*).

Squash blossoms are widely used in American Indian and Mexican cookery. Here is a traditional recipe from the Zuñī Indians of New Mexico.

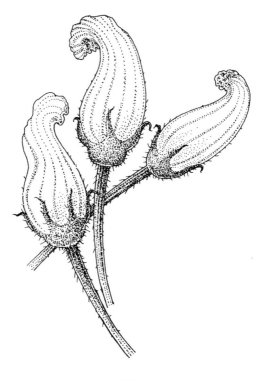

FRIED SQUASH BLOSSOMS

8 fl oz milk	250 ml	1 cup
2 tablespoons flour	2 tbsp	2 tbsp
1 teaspoon salt	1 tsp	1 tsp
½ teaspoon freshly ground black pepper	½ tsp	½ tsp
4 fl oz sunflower or safflower oil	125 ml	½ cup
About 32 squash blossoms		
1 tablespoon paprika	1 tbsp	1 tbsp

To make the batter, combine the milk, salt and pepper in a blender or food processor, Place the squash blossoms in a shallow bowl or metal dish and pour the batter over them, coating them evenly.

In a frying-pan (skillet) heat the oil to 375 °F (190 °C) until a 1-inch (2.5 cm) cube of bread will brown in 60 seconds. Drop the squash blossoms in and turn them frequently so that they turn brown evenly. Leave them to drain on paper towels. Sprinkle them with paprika.

8 servings

$\mathcal{S}$wamp cabbage *Ipomoea aquatica*

Ung choi
Water spinach

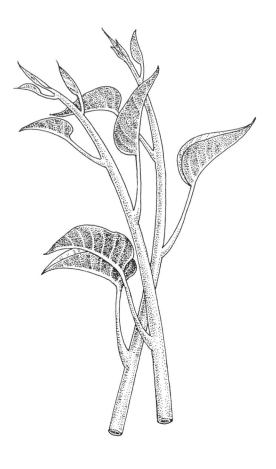

As can be seen from the Latin name, this is not a cabbage at all but an aquatic relative of the sweet potato. It probably originates from Africa, and is grown throughout southeast Asia, but has become popular in Australia and especially in Florida. It has long spear-shaped leaves with cylindrical stems. These hollow stems have little dark bosses on them.

Choose undamaged glossy leaves, and wash them carefully as insects can crawl inside the stems. Swamp cabbage is available throughout the year.

The swamp cabbage has a very delicate flavour. It can be stir-fried and used as an omelette filling, or spiced with nutmeg and almonds and served as a spinach-type vegetable.

Sweet potato *Ipomoea batatas*

Kumara (New Zealand)
Louisiana yam
Yam
Yellow yam

The sweet potato is a tuberous root from tropical America. It is distantly related to the potato, but will only grow in a much warmer climate. The most common forms are long, rather than rounded, and have red skins. The flesh is either whitish, turning slightly violet when cooked, or yellow. The yellow-fleshed variety is confusingly referred to as a yam or Louisiana yam, mainly in the United States, although the yam is a different plant altogether (see YAM).

Sweet potato produces very attractive leaves, which are inedible, and the tubers can be cultivated in cold climates as a houseplant, but will rot immediately if grown outdoors. Always choose firm, unblemished specimens, and watch out for shrivelling.

Sweet potatoes have flesh that is more solid and packed than the ordinary potato, so they need much longer cooking. However, this also means that they are more filling. Due to their sweetness, they can be combined with both sweet and savoury dishes. However they are to be prepared, it is best first to parboil sweet potatoes whole in their skin, in order to ensure that they cook through thoroughly.

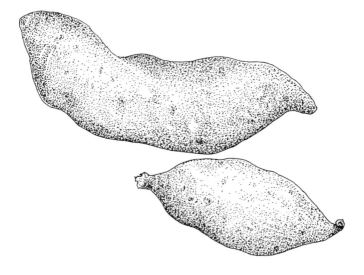

SWEET POTATO AND APPLE DESSERT

1 lb sweet potatoes	500 g	1 lb
5 oz brown sugar	150 g	⅔ cup
4 cooking apples, cored and thinly sliced		
½ teaspoon salt	½ tsp	½ tsp
½ teaspoon cinnamon	½ tsp	½ tsp
2 oz butter	50 g	¼ cup

Parboil the sweet potatoes whole in plenty of water to cover for 45 minutes. Peel and slice them into rings about ½ inch (2 cm) thick. Butter a baking dish and arrange slices of sweet potato in the bottom. Add a layer of apples, then sprinkle with brown sugar. Add another layer of potatoes, then the sugar, and so on, until the ingredients are used up, ending with a layer of brown sugar. Sprinkle with salt and cinnamon. Cut the butter into small pieces and dot them over the mixture. Bake the dish in a preheated 350 °F (180 °C/Gas Mark 4) oven. Serve hot with ice cream or cream.

8 servings

$\mathcal{S}wiss\ chard$ *Beta vulgaris* var. *flavorescens*

Chard
Silverbeet (U.S.A.)

These are the leaves of a variety of beetroot (beet), but all the growth is in the leaves, not in the bulb. The leaves are oval, and are dark glossy green with a thick central rib. This rib is either white or red, depending on the variety. The width of the rib also depends on the variety; some are almost as wide as the leaves, others quite narrow.

Swiss chard grows best in sub-tropical climates, especially the Mediterranean, where, like spinach, it originated. The leaves and ribs are either cooked and eaten separately, or they are all chopped up together. Swiss chard has a similar nutritional content to spinach, being high is nutrients but also containing a lot of oxalic acid. The flavour is milder and sweeter than spinach, and for this reason many people prefer it.

Swiss chard is in season from early to late summer. Choose firm, glossy leaves and avoid those that look limp or wilted.

Spinach beet (*Beta vulgaris* var. *vulgaris*) is another variety of Swiss chard which more closely resembles spinach, although the leaves are not quite as spade-shaped. It can be treated exactly like spinach.

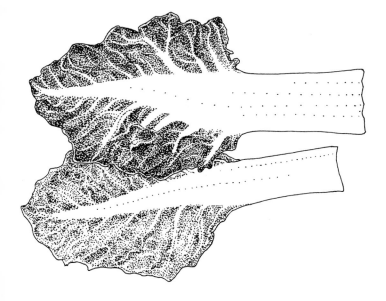

107

BAKED SWISS CHARD

1½ Swiss chard	750 g	1½ lb
2 slices stale white bread, crusts removed, soaked in 4 tablespoons milk		
1 oz dried mushrooms (not Japanese dried mushrooms), soaked in warm water for 20 minutes, drained and chopped	25 g	⅛ cup
2 leeks, trimmed and finely chopped		
3 tablespoons chopped parsley	3 tbsp	3 tbsp
a few celery leaves, finely chopped		
1 garlic clove, crushed		
2 oz grated Parmesan cheese	50 g	4 tbsp
½ teaspoon salt	½ tsp	½ tsp
½ teaspoon pepper	½ tsp	½ tsp
½ teaspoon nutmeg	½ tsp	½ tsp
4 eggs		
3 tablespoons olive oil	3 tbsp	3 tbsp
1 oz fresh breadcrumbs	25 g	2 tbsp

Throw the whole Swiss chard leaves into a large pot of boiling, salted water. Cook for 5 minutes. Drain thoroughly, then chop the Swiss chard into small pieces. Add the chard to the milk-soaked bread, with the mushrooms, leeks, parsley, celery leaves, garlic and half the grated cheese. Season with salt, pepper and half the nutmeg. Stir well to mix. Add the eggs, one at a time, to the mixture, stirring well after each addition.

Heat the oven to 350 °F (180 °C/Gas Mark 4). Oil an ovenproof serving dish generously. Mix any remaining oil with the fresh breadcrumbs. Turn the Swiss chard mixture into the baking dish and top it with the oiled breadcrumbs. Sprinkle with the rest of the cheese and the rest of the nutmeg. Bake for 30 minutes, or until the top is bubbling and well-browned.

4 servings

Tannia *Xanthosoma sagittifolium*

New cocoyam
Pomtannia
Yautia

This plant, with its yam-like tuber, is also grown for its arrow-shaped leaves on very long stalks. The tuber consists of a hairy brown skin over white flesh, like a cocoyam, and can be cooked in exactly the same way. The tubers weigh 6–10 lbs (3–5 kg). The leaves are often sold separately from the tubers, on their long stems, which can be boiled and eaten like SWISS CHARD ribs. The leaves wither rapidly, and should be eaten soon after purchase. They are a variety of CALLALOO.

Tannia originates from India and southeast Asia, but because of its usefulness it is cultivated on lowland soils throughout the tropics. It is particularly popular in the West Indies. The name is of Dutch origin, as is this dish from the Netherlands Antilles.

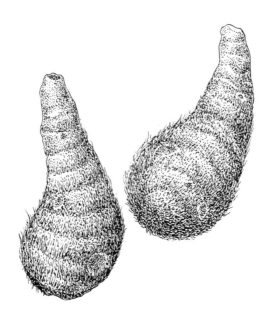

1 large chicken, jointed		
4 oz margarine	125 g	½ cup
salt and pepper		
4 tannias	2 kg	4 lb
8 fl oz Seville orange or grapefruit juice	250 ml	1 cup
6 tomatoes		
1 tablespoon tomato purée (paste)	1 tbsp	1 tbsp
3 medium onions		
8 fl oz strong chicken stock (broth)	250 ml	1 cup
½ tablespoon sugar	½ tbsp	½ tbsp
¼ teaspoon nutmeg	¼ tsp	¼ tsp
6 stalks celery		
1 Scotch bonnet or other small pepper, left whole		

Melt half the butter and sauté the chicken pieces in a large frying-pan (skillet). Sprinkle with salt and pepper while cooking. Remove from the pan and reserve. Peel and grate the tannias and mix them with the Seville orange or grapefruit juice.

When the chicken is cool enough to handle (it should be cooked through), remove and discard the bones, keeping the flesh as whole as possible. Chop two of the onions and the three of the tomatoes and put them in the pan in which the chicken was cooked. Add the tomato purée (paste) and the pepper and cook on a low heat, stirring well, for 5 minutes. Add the stock and continue cooking and stirring for another 5 minutes. Pour this mixture into the grated tannia mixture, scraping the bottom of the pan. Add the sugar and nutmeg.

Preheat the oven to 400 °F (200 °C/Gas Mark 6). Butter and large rectangular glass baking dish. Cover the bottom with two thirds of the tannia mixture, then arrange the chicken pieces on top. Chop the remaining onion and tomatoes and mix. Sprinkle this over the meat. Cover the rest of the tannia mixture. Dot with the rest of the butter. Bake the pom for 45 minutes or until a knife inserted in the centre cuts through it easily. Serve with steamed tannia leaves and boiled stalks, cut into short lengths.

6 servings

Tindoori *Trichosanthes dioeca*

Ivy gourd
Parwal
Tindola
Tindori

These tiny members of the squash family are eaten exclusively by the people from the Indian sub-continent. They look like very small gherkins, but are fatter than these miniature varieties of cucumber. Tindooris are used in curries like other types of gourd. They are light or dark green. Some varieties are slightly bitter, but as in the BITTER GOURD, this flavour is masked when served in a hot curry.

TINDOORI CURRY

4 oz chick-peas, soaked for 2 hours, drained	125 g	½ cup
8 oz tindooris	250 g	1 cup
½ teaspoon turmeric	½ tsp	½ tsp
½ teaspoon garam masala	½ tsp	½ tsp
½ teaspoon salt	½ tsp	½ tsp
1 small green mango, sliced		
1 tablespoon margarine	1 tbsp	1 tbsp
1 small onion, chopped		
1 small green chilli pepper, chopped		

Put the chick-peas into a pot and add 1 pint (600 ml/2¼ cups) water. Bring back to the boil and partially cover the pot. Cook at a slow boil for 30 minutes. Add the tindooris, turmeric, garam masala, salt and mango. Cook for another 30 minutes or until the chick-peas are tender.

Heat the ghee in a frying-pan (skillet) and sauté the onion until it starts to colour. Stir in the chopped chilli pepper and cook for another 5 minutes. Add this mixture to the chick-pea mixture. Serve with rice or Indian bread (naan).

4 servings

Tomatillo *Physalis ixocarpa*

Tomato de Cáscara
Tomato verde

This Mexican fruit, which is used as a vegetable, is not a tomato at all but a relative of the physalis or cape gooseberry (see *Exotic Fruit A–Z*). It resembles a green tomato, with a papery, leafy husk. It is used quite a lot in Mexican cookery, and is sold canned when out of season, but green tomatoes can be substituted for tomatillos if they are not available, though the flavour will not be the same.

The tomatillo is always eaten cooked and has faintly apple-like flavour, which develops during stewing. To prepare the tomatillos for cooking, remove the husks, rinse the fruits and place them in a heavy saucepan. Barely cover with water and simmer, covered, for 10 minutes. Do not cook longer, or they will burst and the flavour of the tomatillos will be diluted with the cooking liquid.

Tomatillos are the main ingredient in the Mexican spicy green sauce known as *salsa verde* or *salsa cruda*.

SALSA VERDE

2 *large red hot chilli peppers* (chile Serrano *or* chile California), *fresh or dried, seeded and chopped*		
2 *tablespoons chopped onion*	2 tbsp	2 tbsp
1 *garlic clove, peeled*		
2 *tablespoons fresh chopped fresh coriander (cilantro)*	2 tbsp	2 tbsp
¼ *teaspoon salt*	¼ tsp	¼ tsp
8 *oz tomatillos, chopped*	250 g	1 cup
⅛ *teaspoon sugar*	⅛ tsp	⅛ tsp

Grind all the ingredients in a pestle and mortar or food processor. When the mixture is well blended (do not make it too smooth; it is best left slightly lumpy) beat in 3 tablespoons of water; the mixture should be fairly runny.

Refrigerate until required but serve as a dip on the day it is made. In Mexico, it is served with all meat dishes and with hot tortillas, tostadas or chilli, but it is equally good as a dip for raw vegetables.

Dieter's Tip   Replace the sugar with a tiny pinch of artificial sweetener. Very healthy, especially if eaten immediately after making.

Water chestnut *Eleocharis dulcis*

Although water chestnuts are more frequently found canned, they can be bought fresh in Chinese shops in the winter. They look like gladioli bulbs, with fibrous skins, enclosing the sweet, firm flesh. Do not worry if they are very dirty, for they clean up well. Remove the skins. Avoid any that are damaged or sprouted.

Water chestnuts are used in Chinese and Thai cooking in a number of dim-sum dishes and are popular in the U.S. as rumaki, a cocktail snack consisting of water chestnuts wrapped in bacon and dipped in a savoury sauce. However, the flavour of raw water chestnuts is unique, and if you can find them fresh, they make a wonderful starter to a meal with a salad of other unusual leaves, such as RADICCHIO.

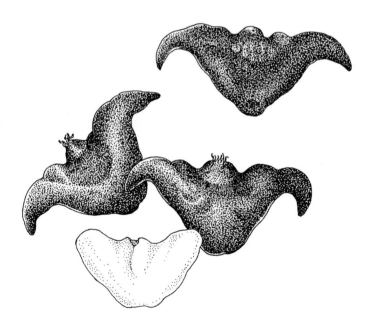

WATER CHESTNUT AND ONION DIP

4 oz salted cashew nuts	125 g	½ cup
¼ pint unsweetened apple juice	150 ml	⅔ cup
1 medium onion		
8 oz water chestnuts, canned or fresh	250 g	1 cup
2 tablespoons finely chopped coriander (cilantro) leaves	2 tbsp	2 tbsp
1 teaspoon soy sauce	1 tsp	1 tsp

Put the nuts and apple juice in a blender and blend into a smooth paste. Chop the onion and water chestnuts very finely and stir them into the paste. Stir in the coriander and soy sauce. Serve chilled with prawn crackers.

8 servings

Wild rice *Zizania aquatica*

Indian rice
Tuscarora rice

Wild rice is not a rice at all, but the grains of a type of rush which grows exclusively in North America, in the lakes of the northern United States and southern Canada. In both countries, it is grown and harvested by Chippewa Indians, who go out to gather it in flat-bottomed boats. It is now also being cultivated in northern California. It is extremely expensive even now that cultivation has been put on a more commercial footing.

Wild rice looks a little like rice with black-and-white grains. It has a deliciously nutty flavour, and is more economical to use than rice, as 8 oz (250 g/1 cup) will serve between four and six people. It should be stored in a cool, dry place, but will not keep indefinitely like rice but only for a couple of months. Leftover cooked wild rice can be kept in the refrigerator for up to five days.

Wild rice can be extended by mixing it with real rice, and it blends extremely well with pecan rice, a new nutty-flavoured rice variety developed in the United States. To cook wild rice, throw it into plenty of boiling water and cook uncovered, so that it 'dances' like rice, for 15 minutes. It goes particularly well with mushrooms (fresh or dried) and game, which is the way the Indians themselves eat it (though they also eat it as a porridge for breakfast). It can also be served cold with a salad.

117

Yam Dioscorea sp.

There are about ten varieties of yam, which originate from all over the tropics. Only one species, the **cush-cush yam** (*Dioscorea trifida*), is native to the New World, though it is very popular in Africa. It has a reddish skin with occasional hairs and it is sometimes scaly or peeling.

All yams are large root vegetables, with white, yellow or purplish flesh, covered with a tightly-adhering thin skin, which must be removed before cooking. All these tubers can be boiled or sliced and fried in the same way as DASHEEN, TANNIA and SWEET POTATO. Like those other vegetables, yams have denser flesh than the ordinary potato, so they need longer cooking. It also means that when they are included in slow-cooking stews, they do not fall to pieces like potatoes, but keep their shape and taste better for being cooked slowly. Another good way to cook yam is to scrape or peel off the skin, rub it with oil or butter and prick it in several places to allow steam to escape during cooking. It can then be baked in a 350°F (180 °C/Gas mark 4) oven for 1½ to 2 hours, or until a skewer will slide smoothly into the middle. Split it open and eat it with salt, pepper and butter. Small pieces of yam cook well in the microwave oven. Put them in a microwave-proof dish and cover with clingfilm (plastic wrap).

Among the most commonly seen varieties of yam are the **Chinese yam** (*D. esculenta*), whose flesh is white and skin scaly and hairy like the cocoyam, the **potato yam** or **aerial yam** (*D. bulbifera*) from tropical Africa and Asia,

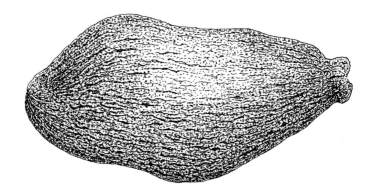

118

whose tubers are exposed and which resembles a small, orange sweet potato, and the **winged yam** (*D. alata*), a southeast Asian variety. The winged yam is very large, and can weigh up to 20 lb (9 kg). The **Guinea yam** (*D. rotundata*), which is distinctive for its yellow flesh, weighs up to half that; it is the yam most frequently grown in the southern United States, and is popular in West Africa and the West Indies.

In tropical Africa, yams are boiled or steamed and served with a rich sauce called *foufou* in Ghana and Nigeria and *foutou* in the Central African Republic and the Ivory Coast. The sauces vary depending on the region and country. Here is a recipe typical of the Ivory Coast.

FOUTOU À L'IVOIRIENNE

2½ pounds Chinese or winged yam	1 kg	2½ lb
2 onions		
4 tablespoons palm oil (or other cooking oil)	4 tbsp	4 tbsp
8 oz cubed stewing steak, fat removed	250 g	1 cup
8 oz eggplant, peeled and sliced	250 g	1 cup
1 small green cabbage, shredded		
1 lb ripe tomatoes, chopped	500 g	1 lb
8 oz cooked chick-peas (garbanzos)	250 g	1 cup
4 oz hot red peppers, seeded and chopped	125 g	½ cup
1 carrot, sliced		
½ teaspoon salt	½ tsp	½ tsp

Peel the yams, cut them into chunks, and put them into plenty of boiling water to cover. Bring back to the boil, then simmer for at least 1 hour.

Meanwhile, grate the onion. Heat the oil in a shallow pan with a lid, and add the grated onion and meat. Sauté until the meat is browned all over, turning occasionally. Then cover with water, and cook on fairly high heat. Add the eggplant, cabbage, tomatoes and chick-peas, red peppers and carrot. Reduce the heat and simmer until the vegetables are tender and the meat cooked through, about 30 minutes. Taste for seasoning.

Drain the yam and mash it by grinding in a food processor (add milk or water if it will not mash well). Put it in a serving dish and pour some of the sauce over it. Serve the rest on the side.

6 servings

SALADE À LA CRÉOLE

1 lb guinea yam or cush-cush	500 g	1 lb
8 oz carrots	250 g	1 cup
1 beetroot (beet), precooked or raw		
1 teaspoon salt	1 tsp	1 tsp
12 oz white cabbage, coarsely chopped	350 g	1½ cups
1 tablespoon chopped green pepper	1 tbsp	1 tbsp
1 tablespoon chopped onion	1 tbsp	1 tbsp
4 tablespoons yoghurt	4 tbsp	4 tbsp
1 teaspoon dried mixed herbs	1 tsp	1 tsp
2 hard-boiled eggs, quartered		
1 teaspoon Tabasco sauce or piri-piri	1 tsp	1 tsp

Scrub the yam and carrots, and the beetroot if raw. Do not peel. Put all of them into a large pan of boiling water, and add the salt. Boil for 45 minutes, or until done.

Drain the vegetables and peel them. Cut them into dice. Finely slice the cabbage and mix it with the chopped pepper and onion. Combine the yoghurt and mixed herbs and toss the chopped vegetables in it. Garnish with the eggs and sprinkle with Tabasco sauce or piri-piri.

8 servings

Yard-long bean *Dolichos sesquipedalis*

Asparagus bean

This is an Asian vegetable which does indeed grow to 3 feet (1 metre) in length if left to itself, but is usually sold when it is only 16 inches (40 cm) long. It is cultivated in tropical and sub-tropical climates, such as California. It has a short season in high summer.

Choose firm, full beans whose ends are unshrivelled. The yard-long bean is so long that one bean is easily enough for two people. Slice it into short lengths and cook it like other green beans.

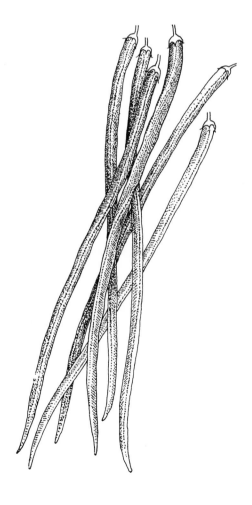

INDEX

Main entries are indicated by **bold** numerals
Photographs are indicated by *italics*

123

124